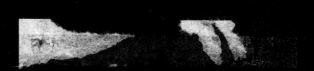

TYCHO & KEPLER

Also by Kitty Ferguson for Walker & Company

Measuring the Universe

Other books by Kitty Ferguson
*The Fire in the Equations: Science, Religion, and
the Search for God*
Prisons of Light: Black Holes
Stephen Hawking: Quest for a Theory of Everything

TYCHO &
KEPLER

The Unlikely Partnership That Forever Changed
Our Understanding of the Heavens

KITTY FERGUSON

WALKER & COMPANY

New York

To my sister, Ginger

First published in the United States of America in 2002
by Walker Publishing Company, Inc.

Published simultaneously in Canada by Fitzhenry and Whiteside,
Markham, Ontario L3R 4T8

For information about permission to reproduce selections
from this book, write to Permissions, Walker & Company,
435 Hudson Street, New York, New York 10014

Library of Congress Cataloging-in-Publication Data
Ferguson, Kitty.
Tycho & Kepler : the unlikely partnership that forever changed
our understanding of the heavens / Kitty Ferguson.
p. cm.
Includes bibliographical references and index.
ISBN 0-8027-1390-4 (alk. paper)
1. Brahe, Tycho, 1546–1601.   2. Kepler, Johannes, 1571–1630.
3. Astronomers—Denmark—Biography.   4. Astronomers—Germany—Biography.
5. Kepler's laws.   6. Planetary theory.   I. Title: Tycho and Kepler.   II. Title.
QB36.B8 F47   2002
520'.92'24—dc21
[B]   2002027445

Visit Walker & Company's Web site at www.walkerbooks.com

Book design by Katy Riegel

Printed in the United States of America

2   4   6   8   10   9   7   5   3   1

CONTENTS

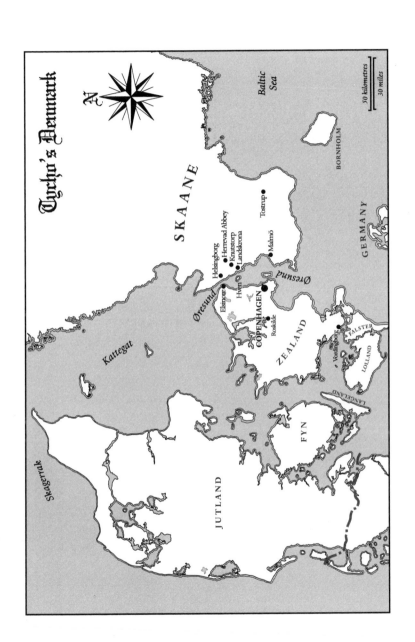

Tycho's Denmark

Tycho and Kepler's Europe

Rostock
Hamburg
Elbe
Magdeburg
Saale
Wittenberg
Sagan
Leipzig
Kassel
Dresden
Benatky
Frankfurt
Prague
Nuremberg
Regensburg
Weil der Stadt
Ingolstadt
Tübingen
Augsburg
Linz
Vienna
Freiburg
Basel
Innsbruck
Graz

N

Venice

150 kilometres
80 miles

ACKNOWLEDGMENTS

The author wishes to express her heartfelt thanks to Owen Gingerich, whose scientific, historical, and bibliographical expertise is exceeded only by his patience, for reading the manuscript and offering corrections and advice; to Sir Brian Pippard, who offered extremely welcome insights on the work of Kepler; to Henrik Wachtmeister, whose family has owned Knutstorps Borg since 1771, and who, after we met him unexpectedly in the churchyard at Kågeröd, graciously welcomed me and my husband and daughter to his home and offered me the use of an extraordinary seventeenth-century drawing of Knutstorps; to Yvonne Björkquist of the Tycho Brahe Museum on Hven, who showed us the sites of Uraniborg and Stjerneborg; to several helpful women at Benatky, whose names I never learned, who did their best to overcome the language barrier and give us an understanding of the layout and history of the castle; to my husband, Yale, and my daughter, Caitlin, who shared these research adventures, navigated the back roads of Sweden and the Czech Republic, took photographs, and read the manuscript; to Anselm Skuhra, who knew the way through the labyrinth of European interlibrary loans and risked his reputation with the University of Salzburg library when I failed to return books on time; to Justin Stagl, whose expertise in European social history rapidly cleared up many mysteries about calendar changes, the politics of the era, and the various names of the river that runs through Prague; to Karoline Krenn, who translated Kepler's writing for me and whose expertise in the period made her translation all the more discerning and accurate; to Gabriele Erhart, whose knowledge of European libraries, galleries, and museums and computer expertise were invaluable for locating illustrations; to Alena and Petr Hadrava, who helped me reach illustration sources in Prague; to James Voelkel, who helped me locate two difficult-to-find pictures; to my literary agents in Europe and America, who read the manuscript at various stages and offered continual encouragement; and to my editors at Walker & Company in New York and Headline Publishers in London, for their splendid work on this book.

PROLOGUE

ON JANUARY 11, 1600, the carriage of Johann Friedrich Hoffmann, baron of Grünbüchel and Strechau, rumbled out of Graz and took the road north. The baron was a man of great wealth and culture, a member of an elite circle of advisers to Emperor Rudolph II of the Holy Roman Empire. Having fulfilled, for the time being, his occasional duties as a member of the Styrian Diet in the Austrian provincial capital, he was returning to court in Prague, the glittering hub of European political and intellectual life.

Among Hoffmann's acquaintances in Graz was a man considerably beneath his own station in society, an earnest young schoolmaster and mathematician named Johannes Kepler. Hoffmann was impressed by Kepler's intense interest in astronomy, an interest he shared, and was aware that Kepler's talents far surpassed those of an obscure provincial teacher. He also knew that Kepler's present situation as a Protestant in Graz was precarious, for the Counter-Reformation was raging there. Only Kepler's position as district mathematician and the expertise he brought to composing annual astrological calendars that predicted everything from crops to wars had prevented his being expelled from the city in 1598 along with other Protestant teachers and ministers.

Hoffmann, for all his lofty status, was a thoughtful, kindly man. It had occurred to him that he might do his youthful friend a service by offering him a ride to Prague in his carriage at no cost to Kepler, who could ill afford such a journey, and an introduction to a far more experienced and distinguished astronomer who had recently arrived there—whose nose was made of gold and silver.

The magnificent Tycho Brahe, the man with the extraordinary nose, was reputedly a difficult person. He had fallen foul of the Danish king and nobility and fled south as a princely refugee with his common-law wife, their six children, wagonloads of fabulous astronomical and alchemy equipment, and three thousand books. Hoffmann's own library was his passion. He was the sort of man to be drawn to an intellect such as Tycho's and also to admire the brilliant networking that had brought Tycho to Emperor Rudolph. In Prague, Tycho Brahe had flowed through the court like fine honey. Rudolph, an eccentric collector of all manner of curiosities, had welcomed him and promised to support him and his learned pursuits in munificent style.

Hoffmann's invitation to ride in his carriage to Prague was a godsend to Johannes Kepler. There was no man in the world whom he so longed to meet as Tycho Brahe. Kepler had already made inept but not entirely fruitless overtures to him. Tycho had had kind, if condescending, words to say about Kepler's first book and had hinted that he would welcome Kepler to join his coterie of assistants. Alas, Tycho Brahe had been too far away in northern Europe and the journey prohibitively expensive. Suddenly, by the grace of God (Kepler had no doubt that God had a direct hand in such matters), Tycho had moved closer, and Kepler had a free ride to Prague. Never mind that it was a one-way ticket, that he had to leave his wife and stepdaughter behind, that Tycho might not be as eager to meet him as he was to meet Tycho, that there was no guarantee a paying job would result. When the baron's carriage drove out of Graz on January 11 and the driver set the horses to the road north, young Kepler was on board.

The journey took ten days, and Tycho Brahe was not in the city when they arrived. He was at Benatky nad Jizerou, a clifftop castle several miles to the northeast. Kepler stayed for a few days as a guest in Hoffmann's palace, considering how best to approach Tycho. Then on January 26, a letter arrived from the great man himself, who had heard that Kepler was in Prague. "You will come not so much as guest," Kepler read, "but as very welcome friend and highly desirable participant and companion in our observations of the heavens." Kepler was apparently not to be just one additional beginner assistant.

Nine days later, on February 4, carrying a glowing letter of introduction from Baron Hoffmann, Kepler rode out of Prague, this time in Tycho Brahe's own carriage with Tycho's son, also named Tycho, and an elegant young man named Franz Tengnagel. They crossed the Labe River at Brandeis, where the emperor had a luxurious hunting lodge, and continued through wooded countryside. The trees, except for the numerous pines, were bare. It wasn't until the next day, February 5, that they reached the first significant change in the landscape, the bluff above the Jizerou River. Poised on top, near the cliff edge, was a square three-story structure of generous but pleasing proportions. It was not the formidable, gloomy fortress some castles were. Perhaps Kepler saw—for it had either been completed then or was in the process of creation—a wonderful fresco covering one entire exterior wall, showing hunting scenes with Emperor Rudolph prominently featured.

Kepler counted among his acquaintances several men such as Hoffmann who were of much greater social stature and wealth than himself. Nevertheless, the lord of Benatky whom he met that day came from a world almost completely outside his previous realm of experience and, in spite of their shared scholarly interests, largely beyond his understanding. Kepler was a well-educated but poorly paid schoolmaster who had spent his childhood in an impoverished, dysfunctional family in small towns on the edge of the Black Forest. By his own description he resembled a little house dog, overeager to please—only occasionally attempting to assert himself by growling or barking and, when he did, succeeding only in causing people to avoid him. Tycho Brahe was renowned throughout Europe as a prince among astronomers and an astronomer among princes; he was supremely well aware of his own superior intellect and status; and he regarded lesser men as just that: lesser men, some of whom he liked and treated well, others not. At the time he met Kepler, he was feeling his age, fifty-three years. Bruised by recent defeats at the hands of enemies in Denmark, he was discouraged about the present state of his work. Nevertheless, he was on his feet again in a different court, lord of an imperial castle, with a promised income from the Holy Roman Emperor that was greater than that drawn by any other man at court. His current public image in Prague was as an elegant and extrovert luminary. Some who had encountered him in different contexts knew he could also be an overbearing, combative, paranoid, and even somewhat malevolent figure.

Though Kepler may have been becoming aware of his own genius, he was still a modest, pious, unassuming, ill-at-ease twenty-eight-year-old in the thrall of a glamorous, formidable, somewhat jaded world figure. Yet in Kepler the mighty Tycho met his match. If Tycho was the dragon of fairy tales, coiled on a fabulous golden hoard—the astronomical observations that he had spent years and a fortune making and now would let almost no one see—then Kepler was the unpromising folklore antihero who was nevertheless endowed with the power to wrest that treasure from him and, from it, forge a new astronomy.

Modern scientists and historians, with hindsight, know this is precisely who Johannes Kepler was. The kindly Hoffmann didn't know; nor did Tycho's son, or Tengnagel. No one at Benatky Castle suspected . . . with the possible exception of Tycho Brahe himself.

Aware only of what Tycho and Kepler had accomplished before that February day when they met, one would not be likely to identify either of them as a prime candidate for immortality. Both men were engaged in developing theories that to modern eyes seem hopelessly misguided. Yet Johannes Kepler and Tycho Brahe would turn out to be two of a mere handful of men who would precipitate humanity into the modern era of scientific inquiry and discovery. When Kepler's exceptional gifts of imagination and inventiveness, his insistence on mathematical rigor and reasonable physical explanations, and his belief that God had created a universe in which harmony and logic prevail came to grips with Tycho's superb, unyielding observational data, the result would be the revelation of profound laws that govern how the heavenly bodies move. Kepler's struggle to find those laws would itself become a prototype for what science would *be* from that time forward. Sir Isaac Newton was referring to Copernicus, Tycho Brahe, Kepler, and Galileo when he said he had stood "on the shoulders of giants."

The colorful, dangerous world in which Tycho and Kepler lived and worked—the courts, universities, cities, palaces, and hovels of Renaissance Europe—afforded them little peace. Against this background, shaped by it and often at its mercy, they nevertheless stood as towering figures not really conformed to any age or time. Nor was their genius the only thing that set them apart. On a more superficial level, they were truly eccentric personalities. Either man, if encountered in a novel, would seem a fantastic or even absurd invention. The same can be said of many of their acquaintances: the unabashedly villainous Nicolaus Bär, the reclusive emperor Rudolph with his largely imaginary royal treasury, the rascal Rosenkrantz who sparked Shakespeare's interest, Kepler's mother Katharina, who was tried for witchcraft. . . .

In this setting and among these people, all manner of events conspired to foil Tycho's and Kepler's loftiest and best-considered plans, destroy their happiness, and distract them from their science. However, this same chain of events brought them together and thus secured for them an immortality that they probably would not have achieved otherwise. No wonder Kepler concluded that the benevolent will of God led men along desperately unwanted paths that only later could be recognized as the right ones.

If invisible cords drew Tycho Brahe and Johannes Kepler over the passage of many years to their crucial encounter, and to the brief, strife-torn, amazingly fruitful relationship that followed it, those strands seem to have been moving almost from the moments of their births.

1

LEGACIES

1546–1561

T H E   O R I G I N   O F the Danish castle Knutstorps Borg—in what was once part of Denmark but is now southwest Sweden—pre-dates written and even oral records. One ancient section of wall in the cellar comes from an eleventh-century structure, but no one knows who lived there then or what the building looked like, secluded among gentle folds of meadow and woodland. There are records from the fourteenth century of an inhabited stone keep. It was probably surrounded by the small lake that appears in sixteenth-century drawings. By that time the keep had become a substantial castle home, the ancestral seat of the noble Brahe family. The lake served as a defensive moat with a causeway and drawbridge.

Here, on December 14, 1546, more than half a century before Johannes Kepler's winter journey to Prague and Benatky, Beate Bille, wife of the Danish knight Otte Brahe, gave birth to twin sons. Only one of them lived, and he was christened Tyge (pronounced "Teeguh"), probably in the small stone parish church of the manor at Kågeröd. Tyge, who would later Latinize his name to Tycho, was Otte and Beate's first son and second living child. His parents did not tell him that he had been a twin.

He was also kept ill-informed about an unusual episode in his early childhood. When Tyge was two years old, his young uncle and aunt, Jørgen Brahe and Inger Oxe, abducted him from his parents' castle and carried him to their own stronghold at Tostrup. As far as records show, and as Tycho Brahe understood this bizarre incident when he was older, there was no outraged protest from his mother or father, no family schism, no scandal, and no attempt to recover him. Otte and Beate by then had a second son, Steen, and were expecting another baby. Tycho would later write simply that his uncle Jørgen "without the knowledge of my parents took me away with him while I was in my earliest youth." It seems that was all Tycho knew.

Short of being a member of the royal family, it was impossible to be higher born than young Tyge.* His ancestors and his relatives had for generations been powerful leaders who served the Danish kings with consummate skill and loyalty, who knew how to maintain a position of influence amid shifting factions at court and how to regain that position if it happened through some stroke of misfortune to be temporarily lost. On the Brahe side the men were warrior knights, at home in the heavy-drinking military circles of the Danish court and ably commanding and administering royal fiefs. Tyge's great-uncle Axel had been one of the first Danish aristocrats to reject Catholicism, so effectively supporting the Lutheran king Christian III during the Reformation in Denmark in the 1530s that he was chosen to carry the scepter at the coronation in 1537. Tyge's father Otte and his uncle and foster-father Jørgen honed their courtly and military skills during that same period of political and religious upheaval. In 1540, six years before Tyge's birth, the king granted them the joint fiefdom of Storekøbing, a step toward increasingly strategic fiefdoms. Otte

---

*Information about Tycho's childhood and youth is found in Tycho's own later accounts and in Victor E. Thoren's splendid scholarly treatment, *The Lord of Uraniborg: A Biography of Tycho Brahe.* Thoren was the leading Tycho Brahe scholar of the twentieth century.

would eventually become governor of Helsingborg Castle, the fortress that guarded the Øresund—the crucial strait that led from the North Sea to the Baltic (see map, Tycho's Denmark, on page xi). He would also hold a seat in the Rigsraad, a body of twenty nobles whose responsibility it was to seat kings, appoint regents, declare war, make treaties, and work with the king on a daily basis in affairs of state.

Tyge's forebears on his mother Beate Bille's side of the family had combined that same kind of secular service to the king with high ecclesiastical positions, and these connections had allowed considerable wealth to be channeled to family members who did not enter the church. When the Reformation came in the 1530s, no fewer than six Billes were in the Rigsraad, and most commanded important castles in Denmark and Norway. Less fortunately, seven of the eight Catholic bishops of Denmark were blood relations, linking the family embarrassingly with what turned out to be the losing side. However, by the time of Tyge's birth the Billes were rapidly repairing their fortunes. The marriage of Beate Bille to Otte Brahe was part of that recovery.

Neither the Billes nor the Brahes were scholars. However, Tyge's childhood abduction had made him the intellectual heir to a third line, the Oxes, the ancestors of his aunt and foster mother Inger. The Oxe family was traditionally more learned and cultivated than most of the Danish nobility. The family had arrived from France at the end of the fourteenth century, and hence their roots did not go back as far in Denmark as the Brahes and Billes. Nevertheless, they had produced four members of the Rigsraad before losing their position in civil upheavals at the time of the Reformation. A decade later, at the time of Tyge's birth, the Oxes' political fortunes were soaring again, thanks to Inger's brilliant eldest brother, Peder Oxe. Inger shared something of Peder's intellectual interests and abilities. She was a woman of great charm, intelligence, and social grace. One of her correspondents and closest friends was the sister of Denmark's King Frederick, Princess Anne, herself a scholar who, despite the time-consuming burden of royal duties and the unlikeliness of the role for a woman of the time, was a skilled alchemist.

Young Tyge was certain to be brought up in a princely fashion, whether he lived with his parents or his uncle and aunt. Having two families gave him an added advantage. He had the attention of an only child in Jørgen's and Inger's castle and would have the support of four younger brothers in Otte's and Beate's family if he later chose to compete in the power politics of the adult world. Most significantly, thanks to the tradition of Inger Oxe's family, Tyge grew up with a somewhat unorthodox view of the world and of the educational and career choices available to him.

There is little record of Tyge's childhood and early youth. Presumably he spent considerable time at his uncle Jørgen's ancestral seat, Tostrup, which was on the side of the province of Skåne that is nearest the Baltic Sea, well to the east of Knutstorp and the Øresund. He also must have visited his parents and his brothers and sisters (seven eventually survived to adulthood) at Knutstorps Borg.

The duties of a Danish knight took Jørgen and almost certainly Inger and their nephew Tyge elsewhere as well. Administering royal fiefs meant periodically spending time in residence, assuring that the buildings and armaments were in a state of repair and ready for defense. However, a nobleman also had to protect his interests at court, networking and second-guessing royal whim. New royal fiefs did not fall one's way if one was continuously absent administering distant royal fiefs, and yet distant royal fiefs did not stay in good enough shape to satisfy the king if one was continually at court. It was a balancing act that required an ambitious vassal to be frequently on the move, and Jørgen clearly carried it off well. Tyge also may have accompanied Inger when she traveled with her own retainers to administer her large share of the Oxe family's domain. When it came to taking an extensive aristocratic household on the road without ever appearing to be a nomad, a young Danish nobleman such as Tyge had plenty of early experience.

In 1552 Jørgen was promoted to the command of Vordingborg Castle, an enormous medieval stronghold on the south coast of Zealand (Sjaelland). Tyge was about six, old enough to participate in some of the pomp and lavish ceremonial entertaining that were part of the life at such an important castle. Vordingborg stood guard over the principal travel route between Copenhagen and the Continent. Duke Ulrich of Mecklenburg and his court rode through the gates in 1556. Princess Elizabeth of Saxony stopped there in 1557 with an escort of no fewer than sixty knights, on the way to visit her grandparents in Denmark. King Christian III himself visited from time to time. Young Tyge was soon no stranger in the company of kings and princes.

Beginning soon after the time of the move to Vordingborg, Tyge began formal schooling. He wrote later that he "was sent to grammar school in [his] seventh year," and he continued elementary studies until about the age of twelve. Tyge's was probably a cathedral school near the castle. At such establishments, sons of the nobility studied side by side with lower-class schoolboys. The curriculum was mostly Latin grammar and religion, with some music and theater, and perhaps Greek and elementary mathematics. Tyge's father thought Latin was a waste of time, but his uncle Jørgen disagreed and insisted Tyge learn it.

When a nobleman's son attended grammar school, he usually lodged in the household of a bishop or other highly placed clergyman so that he could continue to develop, as much as possible outside the castle, the niceties of a gentleman. The days when bishops in Denmark were Catholic aristocrats had ended with the Reformation. In Tyge's school days they were Lutherans of middle-class background, often with large families of their own and not wealthy enough to support luxurious establishments. Nearly all these men had studied at Wittenberg in Germany, and their households emulated those where they had boarded with professors like Martin Luther and Philipp Melanchthon.

Family, boarding students, guests, and colleagues gathered for meals at long tables in a wood-paneled room. Tyge wouldn't have found that too different from mealtimes at the castle except for the conversation. In the household where he lodged, he probably heard for the first time the lively, wide-ranging, intellectual mealtime discussions that traditionally went on around a scholar's table. Mealtime conversation at his uncle's castle, by contrast, would more likely have dealt with warfare, politics, and court gossip.

Tyge Brahe went farther from home to continue his education at the University of Copenhagen when he was twelve—which was not an early age to begin university in those days. He may actually have matriculated, for he recorded the date he began, April 19, 1559. Matriculation was an unusual step for a nobleman's son, because young aristocrats didn't need university degrees as credentials. It was more common for them only to attend selected series of lectures as part of a course of study set by the professor under whose supervision they lived and worked, with no more formal arrangement.

University students lodged with professors rather than bishops or clergymen. Living conditions were comfortable, at least by sixteenth-century standards, because a university appointment, though it did not make a man a member of the nobility, paid fairly well. The professor who provided the lodging also supervised his students' reading and lecture attendance and arranged tutoring with older students residing in the same household. On a smaller, more personal scale, such a household was not unlike a college at Cambridge or Oxford.

The University of Copenhagen was one of the premier universities of Europe. King Christian III, at whose coronation Tyge's great-uncle had carried the scepter, had set the university on a sound financial footing, and Frederick II, the present king, had enlarged its endowment, ensuring an income from landed estates, tithes, and church properties. Among other benefits, the university had the curious right to every eighth swine grazing in the university forests, which perhaps helped to supply the long tables in the professors' households.

The quality of a student's education depended heavily on whose household he belonged to, and there is no record of where Tyge lodged. His uncle and aunt may have placed him in the household of Nicolaus Scavenius, a professor of mathematics, for Tyge's mathematical interests began early, and Scavenius was a client of the Oxe family. On the other hand, he may have lodged in the establishment of Niels Hemmingsen, a renowned professor of theology who would play an interesting walk-on role later in Tycho's life. Anders Sørensen Vedel, who would accompany Tyge on educational journeys abroad, lodged there. Tyge studied Greek and possibly some Hebrew and acquired a classical education with skills in logic, rhetoric, debate, and public speaking, all traditionally considered useful for a young man who intended to follow his forebears into the ranks of the ruling elite.

However, education at Lutheran universities such as Copenhagen went beyond these subjects, largely thanks to Martin Luther's influential follower and friend Philipp Melanchthon. Melanchthon believed that the church could succeed in its mission to teach the path to salvation only if it made education a priority and produced a clergy of scholars strongly grounded in the "liberal arts": In order to understand the Scriptures and the writings of the church fathers, one had to have Latin, Greek, and Hebrew. Knowledge of literature and history lent authority to preaching, which benefited even more directly from mastery of rhetoric and dialectic. Thorough comprehension of both the secular and sacred realms called for arithmetic and geometry. Besides their practical applications, these also helped one understand astronomy, which was considered to be the most heav-

enly of the sciences. Every Lutheran university had at least one pro-
fessorial chair in these mathematical disciplines. Astronomy established the calendar of the church and opened one to the inspiration
of nature and the mind of the Creator, but it also had its practical use
as a basis for astrology. The two subjects were in fact not separate
then, and one of the primary motives for practicing astronomy and
training astronomers was to improve horoscopes. Though his mentor Martin Luther scoffed at such ideas, Melanchthon, along with
many other educated people, thought the fate of human beings was
closely linked to the stars and planets. Most significantly for Tycho
Brahe and, twenty-five years later, for Johannes Kepler, the Philippist
university curriculum promulgated by Philipp Melanchthon also
embodied the humanist ideal that one could not truly comprehend
and master any part of all this knowledge unless one comprehended
and mastered the whole of it.

It was in the atmosphere of such broad intellectual ambitions that
Tyge's interest in astronomy took shape. An eclipse of the Moon on
August 21, 1560, that he either witnessed or heard about when he
was thirteen years old, set fire to his already considerable fascination
with the subject. *

A surviving list of books Tyge purchased provides some information about the astronomy he studied. The books included Johannes
de Sacrobosco's *On the Spheres,* the preeminent introductory astronomy text of the Middle Ages, which Professor Scavenius used in his
lectures; Peter Apian's *Cosmography,* a more advanced book; Johann
Regiomontanus's *Trigonometry;* and an ephemeris (a table showing
the positions of heavenly bodies on a number of dates in a regular sequence) from Stadius. Tyge inscribed his name and the date of purchase, "Anno 1561," in the Apian *Cosmography,* using a Latinized
form of Tyge—Tycho. Tycho spelled the name sometimes with *ij,*

---

*The significance of this event for Tycho was recorded by his first biographer, the early-seven-
teenth-century astronomer Pierre Gassendi.

sometimes with *ÿ,* never as Taecho, indicating that he intended it to be pronounced Teeko, or (closer to the Danish pronunciation) as though the *y* were a German *ü.*

༄

ASTRONOMERS IN THE ERA when Tycho lived thought of their subject as being separated into two parts, described as the *primum* and *secundum mobile.* The *primum* dealt with the way the celestial sphere as a whole "rose" and "set" every night, and the fact that the particular portion of that celestial sphere visible at night changes throughout the year in a regular annual cycle. One needed trigonometry, the most advanced form of mathematics then known, to understand these phenomena in detail, so classroom discussions usually took place on a more general, qualitative level. The *secundum mobile,* involving planetary positions and motions, did require trigonometry.

Typical study of the *secundum mobile* began with Euclid's *Geometry,* a work that had endured since around 300 B.C. (Euclidean geometry is still taught in basic geometry classes.) From there the course went on to trigonometry and planetary theory. In Tycho's university years, planetary theory still meant theory according to Ptolemaic astronomy.

When Greek and Alexandrian scholars such as Aristotle, Hipparchus, and Claudius Ptolemaeus (known as Ptolemy) peered at the night sky, they saw virtually the same panorama that is visible with the naked eye on a clear night now, far enough away from city light. Thirteen centuries after Ptolemy, Copernicus and Tycho Brahe also had no other view than that, for they too lived before the advent of the telescope. Ancient sky-watchers, by scrutinizing the sky with care over long periods of time, had discovered that the motions of the heavenly bodies are not random. Stellar and planetary movement is intricate, but it was possible to calculate well in advance what paths these objects would take and where they would be at a future time. Close observers knew early on that though change, chance, and whim seem to be the rule on Earth, the heavens perform a complex but predictable dance. That dichotomy became a key part of the ancient and medieval worldview.

The best way to describe and explain what one observed in the skies with the naked eye was to think of Earth as the center, with everything else moving around it. That concept still works admirably for purposes of navigation. In fact, to think that things might operate differently demands a leap of fancy that would seem ludicrous to anyone not steeped since childhood in Sun-centered astronomy.

Early astronomers knew, however, that there are phenomena that one observes looking at the sky with greater care over a period of time that seem at odds with a system in which Earth is the center and everything else is in motion around it. Rather than decide that these glitches were significant and stubborn enough to require one to discard the Earth-centered view of the universe entirely and look for another, they chose to attempt to explain the glitches, if they could, *within* an Earth-centered system. Ptolemy's success in doing so was one of the most impressive intellectual achievements in history.

Ptolemy did not begin with a tabula rasa in the second century A.D. by gazing up at the night sky as it appeared to him from near the mouth of the Nile at Alexandria. Instead, he drew together the results of centuries of previous speculation and observation and pondered all of this afresh, applying his own superb mathematical talents. The result, set down in his *Almagest* and other works, was a cohesive explanation of the cosmos that endured and dominated Islamic and, later, Western thinking for fourteen centuries. Finally, even as it was rejected, it provided the springboard for Copernican astronomy and all that has followed from that.

Part of the intellectual worldview of the era in which Ptolemy lived was that the actual appearance of things had to be taken into account in trying to figure out what constitutes "reality." To be plausible, an explanation had to "save the appearances," not contradict them. Though in the early seventeenth century, after Tycho's death, some Ptolemaic astronomers refused to look through Galileo's telescope when it seemed to reveal things that contradicted Ptolemy, Ptolemy himself did not ignore "what can be seen up there" in favor of some mathematical fable. He would have looked through Galileo's telescope. However, nothing about the appearance of the heavens, as Ptolemy and his predecessors were able to study them, forced them or him to reject the intellectual tradition that held that all heavenly movement occurred in perfect circles and spheres.

The "spheres" were not the planets themselves, but transparent glass spheres in which the planets traveled. Astronomers spoke of "crystalline" spheres, each having an inner and an outer wall, with space between the two walls for the planet to move. The spheres were nested one within the other, with each successive sphere just small enough to fit within the one outside it. They were tightly packed with no extra space left between them, but not so tightly as to prevent their moving, one against the other, with the outer surface of one sphere scraping against the inner surface of the next larger.

Sitting at the center of this system of nested crystalline spheres was Earth. The outermost sphere in the arrangement was the sphere of the stars. The innermost sphere—nearest Earth—was the sphere in which the Moon moved. The others each contained a planet, except for the one that contained the Sun. Each body could move only between the outer and inner walls of its own sphere. Not all scholars agreed about the nature and mechanics of these spheres, but there was general agreement that a planet couldn't break through those walls and enter another planet's sphere. In fact, in this system, *no* heavenly body could break through the walls of a sphere. That would shatter it. This last restriction became significant for Tycho Brahe and Johannes Kepler.

One of the most stubborn problems for ancient astronomers was

(a)   (b)

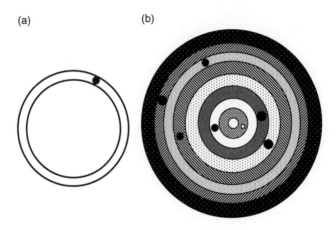

Figure 1.1: Early astronomers thought of the planets and the Sun and Moon as each moving in its own "crystalline" sphere (a), with these spheres nested one within the other and Earth at the center (b).

how to explain a phenomenon known as the "retrograde" movement of the planets. A planet normally moves from west to east against the background of stars. However, during a period known as its "opposition," when it is on the opposite side of Earth from the Sun, a planet appears for a while to move from east to west. Scholars were faced with the problem of explaining this in a model that required uniform movement and perfect circles and spheres. The solution, devised before Ptolemy, was ingenious.

A carousel is a helpful analogy for understanding the idea: On the simplest carousel, the horses are bolted directly to the floor, which is a large, rotating disk. They circle, as the disk rotates, but they have no other motion. If the amusement park is dark and there is a light attached to the head of one horse, an observer, positioned at the center of the carousel in such way as not to move with the rotating disk, sees the light circle steadily. It will have no "retrograde motion"— never seem to back up.

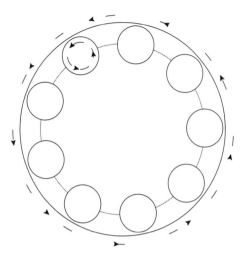

Figure 1.2: On this carousel the horses are attached to smaller rotating disks that ride on the perimeter of the large, rotating floor. As the carousel turns, the horses not only make their way all the way around the large circle but also move in smaller circles, chasing their tails.

Suppose instead that the observer does occasionally see the light stop, back up for a while, and then resume its former motion. This isn't a random occurrence, as it might be if the light were on the cap of the ticket taker as he moves among the riders, or if a large firefly happened to venture into the carousel. The backing up happens regularly and predictably. The observer decides that the horse with the light on its head must not be bolted directly to the rotating disk. Instead, each horse is part of a minicarousel perched near the edge of that disk. Hence, in addition to their motion with the disk, the horses are moving around in smaller circles, chasing their tails.

By the same token, in a stroke of insight, ancient astronomers realized that if the planets moved continually in smaller circles centered on the rim of a larger circle centered on Earth, the result would be the regularly occurring retrograde motion they were observing.

The technical term in Ptolemaic astronomy for the small circle in which a planet moved was *epicycle*. The larger circle on which the epicycles turned (in figure 1.2, the inner circle—the radius at which the minicarousels are bolted to the floor) was the *deferent*. By adjusting the size, direction, and speed of the epicycles, astronomers could explain many irregularities they observed in the way the planets, Sun, and Moon move. A planet traveling in its epicycle would sometimes be closer to Earth and sometimes farther away, which explained apparent variations in its brightness. In Ptolemaic astronomy a planet's sphere was just large enough for the planet to cartwheel along on its epicycles.

Tycho, Kepler, and their peers at university also learned the use of the *eccentric*. With an eccentric, the planet (perhaps simultaneously traveling in an epicycle) orbited Earth, but the orbit wasn't

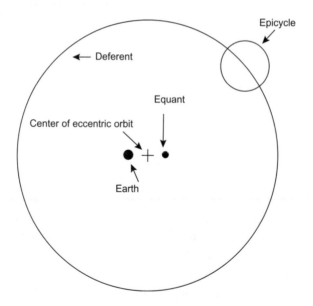

Figure 1.3: Devices of Ptolemaic astronomy: eccentric orbit, deferent, epicycle, and equant.

centered precisely on Earth. Its center was a point a small distance away from Earth.

Epicycles, deferents, and eccentrics were devices Ptolemy refined from earlier astronomy, but another, the *equant*, was probably his own invention. Many astronomers were uncomfortable with it, for it was not only complicated to use but also seemed to cheat a bit on the requirement of uniform motion. The equant was an imaginary point that Ptolemy used to rationalize a planet's apparently slowing down and speeding up as it wheeled in its epicycles around the deferent. It was possible to establish mathematically that if one were able to view the heavens from the equant, the velocity of a planet would *appear* to be uniform, though from Earth or the eccentric center of the orbit it would appear to vary.

Ptolemy combined these devices in a complex and highly success-ful model of heavenly motion. Without removing Earth from its po-sition as unmoving center, his astronomy could, with a surprising de-gree of accuracy, predict and account for the changing positions of the Sun, the Moon, and the five planets that were known at that time. Fulfilling the hopes of centuries of scholars before him, Ptolemy was able to accomplish this feat entirely in terms of circles, spheres, and uniform motion.

Students and scholars of Tycho Brahe's and Johannes Kepler's gen-erations were also steeped from childhood in a worldview that far pre-dated Ptolemy: Nineteen centuries removed from its origin in ancient Greece, Aristotelian philosophy and cosmology still had an enormously strong hold on the thinking of scholarly and religious Europe. This worldview held that everything below the orbit of the Moon was subject to change, degradation, and decay, while the heav-enly spheres beyond the Moon were a realm of unvarying, eternal perfection. Both experience and observation gave weight to these ideas. Before the telescope, there was little evidence to challenge the perfect immutability of the heavenly spheres. Nor was it possible to deny that things were different on Earth.

This dichotomy had entered the thinking of European Latin-speaking scholars when the first Latin translation of Aristotle appeared in the twelfth century. These men knew nothing of Ptolemy, though the heritage from his astronomy was still flourishing in Islamic parts of the world. They came to revere Aristotle, instead, as the final authority on science and cosmology, and Aristotle's cosmology, filtered through the understanding of these scholars, merged with medieval Judeo-Christian thought. Somewhat later, when Ptolemaic astronomy arrived in Latin Europe, there was a second merger. Scholars, who in both instances were all clergymen, put prodigious effort and much debate into reconciling the Bible first with Aristotle and later with Ptolemaic astronomy. To do so, they began to give the Scriptures a less literal, more metaphorical interpretation. What emerged over time was a coherent body of philosophical, scientific, and religious thought, with astronomy giving a visual, geometric structure to abstract medieval Judaism and Christianity. Aristotle's picture of the degraded, changing, decaying nature of Earth and the pristine perfection of everything beyond the Moon was consistent with the Judeo-Christian view of fallen, lost humanity on Earth and the eternal, holy realms above.

From the thirteenth century until the sixteenth, most educated Europeans accepted this worldview as reality. For them, astronomy seemed capable not only of describing and predicting planetary movement and providing a map of the cosmos but also of accurately describing the human condition, with men and women pitifully torn between the passions of the squalid, death-ruled Earth and the lure of the deathless, sacred heavens. In the fourteenth century, Dante gave this worldview eloquent poetic expression in his *Divine Comedy,* describing a journey downward through nine circles of hell toward the center of Earth, the most debased point in the universe, and a journey upward through the celestial spheres of the planets to reach the throne of God. It is no wonder that Melanchthon believed that the study of astronomy was essential to the clergy.

Both young Tyge and the young Kepler found this picture compatible with all they knew of Earth and heaven. However, though both men remained devoutly religious all their lives and found no contradiction between their science and their belief, they would leave this primitive worldview in tatters.

2

ARISTOCRAT BY BIRTH, ASTRONOMER BY NATURE

1562–1571

TYCHO CELEBRATED his fifteenth birthday in December 1561. He had been at university for three years, and it was time to begin a new phase of his education. As the scion of a noble family, presumably destined for public life, he needed to become familiar with the history, music, art, literature, and architecture not only of Denmark but of the whole of Europe. He also needed to be able to speak other languages besides Danish and the Latin, Greek, and Hebrew he had already studied, and to experience the idiosyncrasies of foreign courts and their rulers and learn something of military science and political theory.

Traditionally, a young Danish aristocrat began to acquire this sophistication by serving as a page in the household of a kinsman and then as squire to a foreign nobleman. At the age of twenty-one, he would move up to the level of a courtier or a knight and finally come back to Copenhagen to serve the king and his court, the last stage of training that equipped him and gave him the credentials to take on the governance and defense of a royal fief. Tycho's four younger brothers followed that path. Two of them eventually became members of the Rigsraad. Had Tycho grown up in his father's castle, or had his aunt not been from the intellectually inclined Oxe family, he probably would have done the same.

The paths of the young men in Inger Oxe's family, like those of the Brahe family, had led them abroad, but not to foreign courts and wars. They had gone to foreign universities. Inger's older brother Peder Oxe had traveled for five years with a tutor from university to university, and this alternate form of education had not prevented his succeeding in public life. Indeed, though by the time Tycho left Copenhagen Peder had suffered a severe downturn of political fortune, there had for a while been no man in Denmark whose career had been more spectacularly successful. In 1548, when Tycho was still a child, Peder led the entourage of Princess Anne of Denmark when she wed the duke of Saxony, and that was the start of a meteoric rise to power that brought Peder, at the age of thirty-two, to a position in the Rigsraad.

Jørgen and Inger chose the University of Leipzig, in Saxony, as the place for young Tycho to begin his foreign experience. They had traveled to Saxony themselves for Princess Anne's wedding, and Inger still corresponded regularly with Anne, who was now Electress Anne of Saxony. The language there was that spoken at the Danish court, a "pure" form of High German. Furthermore, Saxony was the birthplace of Lutheranism.

A fifteen-year-old wasn't sent abroad alone. Jørgen and Inger carefully chose Tycho's "preceptor," Anders Sørensen Vedel. Four years older than Tycho, he came from a respectable middle-class background and had excelled at the University of Copenhagen. As a preceptor, Vedel's duties combined those of companion, chaperon, and tutor. In return for the payment of his expenses, and supposedly adhering to instructions in letters sent from home, he was to supervise Tycho's university studies, act as his spiritual and moral guide, see that he received language instruction and lessons in fencing, riding, and

Anders Sørensen Vedel, Tycho's preceptor and lifelong friend, in a 1578 oil painting by Tobias Gemperle.

dancing, and manage the purse. All of these responsibilities added up to a formidable assignment, but being a preceptor was nevertheless a good way for a young man of modest means to support his own education abroad and make invaluable contacts.

Tycho and Vedel left Denmark on February 14, 1562, traveling, for safety and companionship, as part of a caravan. The journey took five weeks, first by ship across the icy Baltic and then on horseback along the roads beside the Elbe and Saale Rivers (see map, Tycho and Kepler's Europe, on page xiii). Many Danish students attended the University of Wittenberg, but Tycho and Vedel's travels took them two days beyond that, to Leipzig.

Germany, then, was still three centuries from unification, and Saxony was not nearly so significant or wealthy a power as Denmark. Leipzig was, however, the site of one of Europe's largest and most important universities. Though there were few Danes in residence, one of Vedel's classmates back home had a brother there, and there were similarities to the University of Copenhagen that made it less alien to Tycho and Vedel. Instruction was in Latin, as it was in all European universities, and education went according to the pattern and philosophy of the Philippists, with many Philippists among the faculty. At Leipzig, Tycho studied classical languages and classical culture. He also continued, at first clandestinely, with astronomy.

Tycho later wrote that he had "bought astronomical books secretly and read them in secret." He studied the constellations from maps drawn by Albrecht Dürer. He began to keep track of the planets by a rough method of lining up a planet and two stars, holding up a "taut piece of string," and then figuring the positions of the planet from the locations of the two stars on a little globe that he owned, "no bigger than a fist." Flawed as this method inevitably was, Tycho was nevertheless able to come to a conclusion that impressed him deeply. Neither the Alfonsine Tables, which had been calculated in the thirteenth century using Ptolemy's Earth-centered model, nor the Prutenic Tables, based on Copernicus's Sun-centered astronomy and drawn up much more recently, were dependable in their predictions of planetary positions.*

When sixteen-year-old Tycho began to keep a logbook of his own astronomical observations, in August 1563, during the second summer of his stay in Leipzig, the first record he made in it was of an observation of Mars, and the second was of a conjunction of the planets Jupiter and Saturn. A conjunction is the coincidence of two or more heavenly bodies at the same celestial longitude. To an earthly observer, one appears to pass the other. This happens with Jupiter and Saturn once every twenty years. Tycho found that neither the predictions of the conjunction based on Ptolemy nor those based on Copernicus were correct. The discrepancies were great enough to show up clearly even with his amateur efforts. The Copernican tables

---

*The Alfonsine Tables, based on Ptolemy, were drawn up in 1252 by fifty astronomers under the patronage of Alfonso X of Castile. The Prutenic Tables, based on Copernicus, were drawn up in Wittenberg by a young colleague of Philipp Melanchthon, Erasmus Reinhold.

weathered the test slightly better than the Ptolemaic ones, which were off by an entire month. The cocky sixteen-year-old concluded that someone ought to produce better tables, and he began to think of himself as the person destined to "rectify this sorry state of affairs."

In Tycho's day a conjunction of the planets was considered to have more than astronomical interest. It was of great astrological significance. For practice, Tycho was casting predictions and horoscopes of famous men (without their knowledge) and recording the results in a notebook.

Despite the best efforts of Vedel to keep his charge on track with his studies of other subjects, Tycho was soon practicing his astronomy more openly. Bartholomew Schultz, a more advanced student at Leipzig, gave him some instruction and introduced him to the more technical side of the subject. Tycho needed a better instrument. He managed to acquire one—his first real astronomical instrument—a cross staff, or radius. On May 1, 1564, when he was seventeen, he entered his first observation from it in his log. After that, he often "stayed awake the whole night through, while my governor [Vedel] slept and knew nothing about it; for I observed the stars through the skylight."

Though Schultz showed him a trick, using "transversal points" that would allow him to obtain more refined measurements (see figure 7.8c), Tycho soon became dissatisfied with the imprecision of his radius. He began to discover errors in his data, which he traced to faulty logic in its construction. The only recourse was to rectify them with a table of corrections, for he "had no opportunity of having a new [instrument] made, since my governor, who held the purse strings, would not allow things of this kind to be made for me." Tycho had already begun to be more seriously concerned about the precision of observations than anyone before him, or any of his contemporaries.

The next December, 1564, on Tycho's eighteenth birthday, one of the professors at dinner described an illiterate craftsman he had met as "an astronomer by nature." Tycho recorded that phrase in his

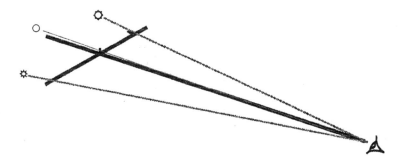

Figure 2.1: The cross staff, or radius: To find the apparent or "angular" distance (see appendix 1) between two stars, an observer sighted from where the "eye" appears in the drawing and slid the crossbar up and down the staff until the distance between the two stars was exactly covered by the length of the crossbar. Tycho's cross staff was more sophisticated. Like the one shown here, it had two sights on the crossbar, and one of them was movable. He adjusted the movable sight so that he could see one star through it and the other through the sight fixed at the center of the crossbar. He then found the angular distance between the two stars by reading the scales etched on the crossbar and the staff and using a table of tangents.

notebook. The same month he set himself a research project to test a popular notion that the day-to-day positions of the heavens during the twelve days of Christmas, beginning on December 25, presaged the month-by-month pattern of the weather for the next twelve months. During those twelve days Tycho recorded every feature he could of the heavens, planning to check the weather throughout the coming year to see whether the data agreed.

The following May, 1565, Tycho and Vedel returned home on a vessel that threaded its way among warring ships. Denmark was once again engaged in combat with its perennial rival, Sweden. Disembarking at Copenhagen, Tycho continued his journey to Knutstorps Borg, taking observations and recording the latitude at each stop on the way.

Knutstorps Borg in May 1565 was much larger and more impres-

sive than it had been when Tycho was born there. In 1551 Otte and
Beate had rebuilt the ancient manor house. From a causeway and
drawbridge over the lake, a gate now led into a square central court
surrounded by four ranges of buildings. Those on the north and
south side had tall, steep roofs with scalloped end gables. The walls
were three to four feet thick, with arrow-slit windows, for the castle
was intended to be a fortress as well as a stately home.

Though Otte and Beate's manor was never directly threatened,
the war with Sweden came close to Knutstorp in the form of border raids. Beate's mother and father, Tycho's grandparents, died defending their castle, Baahus. Even more drastic for Tycho was the
death of his uncle Jørgen in June 1565. Jørgen was vice admiral of
the Danish fleet, which was in Copenhagen for repairs and repro-
visioning. King Frederick and Jørgen had been drinking, and the
king fell into the water under Amager Bridge. Jørgen dived in to
rescue him. The king recovered, but Jørgen either drowned there or
died almost immediately thereafter from injuries or illness attributable to his rescue effort.

His foster father's death should have left Tycho a wealthy man.
Jørgen had no children of his own, and he had been in the process of
making Tycho his legal heir. Unfortunately that process had not been
completed. Inger held life tenancy of the manor at Tostrup as her
"widow's jointure," but eventually, at her death, the castle and the
income from its hundreds of tenants would not come to Tycho but
revert to general distribution among the Brahe family. Meanwhile,
responsibility for Tycho, who at eighteen had not yet reached his majority, fell to his natural mother and father.

Tycho left Denmark again a year later, in the spring of 1566.
Somehow he had convinced his father to allow him to continue his
education abroad rather than take advantage of the excellent opportunity the war provided to begin a career of civil service to the king. This
time Tycho's destination was Wittenberg, where Anders Vedel was already working toward his master of arts degree. Five months after

Tycho arrived in Wittenberg, an epidemic struck the town, and most of the students fled. Tycho moved to the more northern university town of Rostock and began studies there in September. This move, though not particularly significant for Tycho's future as an astronomer, was extremely significant for his future physical appearance.

On December 10, Tycho was a guest at a betrothal celebration. During the dancing, he fell to quarreling with another aristocratic Danish student, his third cousin, Manderup Parsberg. The quarrel may have erupted as the result of some levity, at Tycho's expense, about an unfortunate astrological prediction he had made that autumn. There had been an eclipse of the Moon on October 28 that Tycho concluded presaged the death of Suleiman the Great, the Turkish sultan. With a flair for the dramatic and poetic, Tycho composed a Latin poem announcing this prediction. Then the news arrived that the sultan had already been dead for six months.

Whether or not it was an insult having to do with that embarrassment that precipitated Tycho's and Parsberg's dispute, it did not end at the betrothal celebration. The two young men resumed their argument at another party on December 27.

Two days later, December 29, Tycho's astrological computations told him that there would be some sort of accidental happening. In spite of the fact that one of his predictions of late had been notoriously inaccurate, he decided to take the warning seriously and not go out at all that day. However, when evening fell, he ventured downstairs in his lodging house to supper. Before long he and Parsberg were quarreling again, wrought up, each demanding that the other draw his sword. They rose abruptly from the table and went out into the churchyard.

A woman in the room knew Danish and understood the seriousness of the dispute. She urged other diners to pursue the young men and prevent their damaging or murdering one another. It was too late. The company emerging from the dining room found a bloody scene. A blow from Parsberg's broadsword had cut away a good portion of

Tycho's nose and just missed proving fatal. Later portraits show a diagonal scar across Tycho's forehead and a curving line across the bridge of his nose. Tycho endured a lengthy, painful, and anxious period of convalescence that winter. His doctors couldn't reverse the disfigurement, for skin grafting, though done in other parts of the world, was unknown in Europe until about two decades later.* Fortunately, infection did not set in, and sufficient scar tissue formed.

Although Tycho lost part of his nose at Rostock, he gained two new interests that remained with him for the rest of his life: medicine and alchemy. When he returned to Denmark in April, he was already experimenting with ways to replace his nose artificially. Later descriptions indicate that eventually, with fair success, he made a false nose by blending gold and silver to a flesh color, or used copper for everyday wear. He held the nose in place with an adhesive salve that he always carried with him in a small box.

When Tycho left Denmark for the Continent a third time, the next December (1567), he did so over much stronger protests from his father. The war had offered a superb opportunity to begin a political career, which Tycho had ignored. Now the steady recovery of Peder Oxe's influence made success for Tycho, as foster son of Peder's sister, a certainty, if Tycho would only be persuaded to enter public life. Tycho had come of age when he turned twenty-one earlier that December, but that didn't leave him free to flout his father's will, for he was still dependent on him for financial support. Tycho had just missed inheriting his uncle's fortune, and it was doubtful that he would be able to earn a living in the pursuits he was choosing. Nevertheless, to Otte Brahe's intense frustration, his eldest son turned his back again on a promising future and headed across the Baltic to Rostock.

Though modern popular opinion might have it that in earlier

---

*Tycho's first biographer, Pierre Gassendi, reported that skin grafts for nose replacement (performed by members of the potters' guild) were done routinely and with a good success rate in Tycho's day in India, where adultery was punishable by amputation of the nose.

centuries an aristocrat had unlimited opportunity, while the lower classes were sadly constrained in their career choices, the fact is that in the Europe of Tycho's day a nobleman's son was fairly strictly limited to the career paths that Tycho's forebears and relatives took—knighthood, the administration of fiefs, or government civil service. Tycho enjoyed university life, but there was no future in that, for most scholarly positions in the universities were closed to noblemen.

There was, however, another possibility: a canonry. The cathedrals in Denmark and Norway—having recently become Lutheran—still retained their rich landed endowments. Royal administrators awarded the positions of canons of the chapters of these cathedrals both to government servants and to men of learning. Commoners and noblemen alike were eligible for a canonry, which carried with it the income from the endowment. Becoming a Lutheran canon did not require a man to enter holy orders, live in the cathedral precincts, or assume a less secular lifestyle. It was the ideal solution for someone who wanted to have a career as an astronomer and a scholar while upholding tradition and retaining his dignity as a member of the nobility, without trespassing on the career opportunities of men of another class of society. There was an excellent precedent, for Nicolaus Copernicus had been a canon of a cathedral chapter.

Hence, while Tycho was returning to foreign climes, heedless of the future, his more sober friends and relatives at court, who included the influential Peder Oxe, set to work to procure a canonry for him. On May 14, 1568, royal letters patent designated Tycho to take up the next vacant canonry at Roskilde Cathedral. The position was reserved for him, though he had to wait for an opening. Although he could not know it had happened, there had been a minuscule tightening of the cords that would draw him over the next thirty-two years to that February day at Benatky when Johannes Kepler arrived.

Meanwhile, Tycho's lodgings in the law college at Rostock were providing an excellent setting for astronomical observations, and he was also finding time for his new interest in medical alchemy. But

when university authorities charged him a hefty fine, possibly because of the duel with Parsberg, Tycho left rather than pay it and
traveled south. At Arnstadt there was to be a ceremony in which
Count Günther of Schwarzburg-Rudolstadt presented Tycho's
younger brother Steen with his spurs, a warhorse, and a harness of armor. Steen's training had taken the traditional path, and he was now
well ahead of Tycho on that path.

By September, Tycho's meanderings across Europe had taken him
to Basel, and there he matriculated at yet another university. After a
few months he moved on to Freiburg, where he was impressed by
some celestial models demonstrating planetary motions according to
the theories of Ptolemy and Copernicus.

As Tycho's restlessness and his travels continued, he gradually
came to realize that after nine years of university he had learned all he
could from professors and books. It was time to embark on more independent work, to direct his own life and education, and especially
to experiment with the design and construction of his own observing
instruments. His cross staff, even though it was one of the best available and he had found ways to compensate for some of its deficiencies, had long ago proved inadequate to Tycho's needs.

In spring 1569 Tycho's travels brought him to the fine old imperial city of Augsburg, and he found it so much more congenial than
any of the other cities he had visited that he stayed for fourteen
months. It was in Augsburg that he began to follow through seriously on the plans to improve his instruments.

His first project was to fashion a new pair of compasses.* Such
an instrument was used to measure the angular distance (see appendix 1) between stars, or between a planet or comet and stars
near it in the night sky. Tycho wrote of an earlier, more primitive
version that he used it by "placing the vertex close to my eye and directing one of the legs toward the planet to be observed and the

---

*A pair of compasses, like a pair of scissors, is one instrument, not two.

other toward some fixed star near it." Though the new pair of compasses was a good enough instrument for Tycho to include it in the catalog of his instrumental achievements twenty-five years later, it was not completely satisfactory. One problem was that he could not manage to mount it in such a way as to allow exact, steady sighting. Another was that Tycho had set himself a goal of accuracy to a minute of arc. The smallest division on the compasses was one *degree* of arc; a minute of arc is *one-sixtieth* of a degree of arc. To approach the precision Tycho wanted, it would be necessary to have an instrument large enough to include the device called "transversal points" he had learned about from Bartholomew Schultz in Leipzig. Meanwhile, as he had done earlier for the cross staff, Tycho drew up a table of corrections.

Not long after his arrival in Augsburg, Tycho was lingering in front of a shop, conversing about the challenge of building an instrument of sufficient size, when a wealthy alderman of the city, Paul Hainzel, chanced by, overheard the discussion, and enthusiastically joined in. Hainzel and Tycho soon discovered that they had a connection. Hainzel's brother had once been a fellow student with Peder Oxe. So intrigued did Hainzel become with Tycho's instrument project that he offered to underwrite the cost of it.

Tycho, who by now had a good idea of the instrument he wanted, set to work on the design and engaged the necessary craftsmen. The *quadrans maximus,* or "great quadrant," was completed in a month, and it was the largest instrument Tycho would ever design—indeed, that he would ever see—twenty feet high and requiring forty men to set it in place in the grounds of Hainzel's estate just outside Augsburg. Tycho boasted about its accuracy, but it had problems. Tycho made only one entry in his log each night, perhaps because it required so many of Hainzel's servants to rotate the cumbersome quadrant into the necessary plane and swing the arc up to the appropriate elevation. They had to keep it there while more adjusting brought the planet being observed into the sights. Only then could the numbers where the

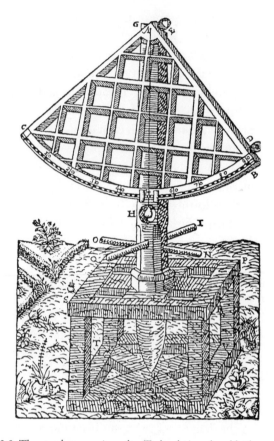

Figure 2.2: The *quadrans maximus* that Tycho designed and had constructed in Paul Hainzel's garden near Augsburg was built mostly of oak, with brass for the graduation strip on the arc and the plumb bob. The length along one edge (from *C* to *G*, for example) measured more than fifteen feet. The entire triangular section hung from its point at the top, allowing it to be swung up until Tycho could see the heavenly body through both the sights (*D* and *E*) at once and note at what number the plumb line (*H*) fell. The drawing is from *Astronomiae Instauratae Mechanica*.

plumb line fell be written down, giving the altitude* of the planet above the horizon. Sleepy retainers, no matter how loyal to Hainzel, could not be expected to perform this operation more than once a night without rebellion.

However, Tycho had been using his new quadrant for less than a month when it proved its worth by bringing him to the attention of the academic elite, in the form of a well-known philosopher and iconoclast, Petrus Ramus. Ramus visited Augsburg in April 1570, and the two men—one a neophyte and the other a renowned scholar—were soon engaged in lively discussion about the future of astronomy. Ramus wanted to rid astronomy of all "hypotheses" and rebuild it totally from observation. The ideas that heavenly bodies must move in circles or epicycles with uniform motion and all other such assumptions would have to be discarded. Tycho agreed that such axioms of physics were not cast in stone, not immutable "truth." He made the distinction, however, that the axioms of geometry *were* based on observation, not some flight of fancy. Furthermore, astronomy would not be possible without any hypotheses at all. He chose to side with the humanists and Philippists: There was recognizable order and harmony to the cosmos, and since there was, it would lend itself to hypothesizing.

Tycho and Ramus were in complete agreement that the future of astronomy lay in numerous and exact observations. The celebrated Ramus, in his next book, described the *quadrans maximus* with admiration and mentioned a young Danish nobleman named Bracheus as its designer.

Tycho had begun a second instrumental project, a large celestial globe made of wood, when a message arrived from his father. Otte Brahe was summoning all his sons back to Denmark, for his health was declining rapidly. Tycho was obliged to leave the globe in the

---

*For the definition of altitude, horizon, and other vocabulary of astronomy, see chapter 7 and appendix 2.

hands of Hainzel and return home. By Christmas 1570 he had arrived at Helsingborg Castle, where his ailing father was still commander. The castle tower was a good vantage point for viewing the heavens, and entries in Tycho's log show that he observed the Moon.

Otte Brahe died the following May, 1571, survived by his wife Beate Bille, seven grown children, and one grandchild. Danish law stipulated that a lifetime widow's jointure be set aside for Beate. The sons and daughters would share the remainder of the inheritance, with each son receiving twice a daughter's share and sons given preference when it came to inheriting manors. Being the eldest son was not an advantage under the Danish inheritance laws: Tycho received no more than each of his brothers.

Tycho and Steen, the next eldest son, jointly inherited the ancestral estate at Knutstorp, splitting the income from two hundred farms, twenty-five cottages, five and a half mills (one mill had been split in an earlier inheritance), and the manorial production and seigneurial rights of the estate. This was a small inheritance by the standards of Danish nobility: Tycho had come within a hairbreadth of inheriting a much larger fortune when his uncle died. Nevertheless, it was clear that when Otte's will was settled, a procedure which normally took several years, Tycho would be financially independent.

As eldest son, his first duty was to take his widowed mother back from the castle at Helsingborg to the family ancestral home at Knutstorp. In the months that followed, much of Tycho's time was devoted to helping her and negotiating the settlement of the estate. Sometime during this period, Beate told Tycho about his twin. The revelation touched him profoundly, and he wrote a poem in Latin, phrased as though it came from the dead child who looked with pity on the one who was living: "He dwells on earth," the brother in the poem says of Tycho, "while I dwell on Olympus."

3

BEHAVIOR UNBECOMING
A NOBLEMAN

1571–1573

IN SPITE OF his steady drift toward the life of a scholar, during
the period following his father's death Tycho spent some of his time
across the Øresund at the royal court in Copenhagen. He already had
powerful family connections there, and now he fashioned closer ties
with King Frederick II himself.

The kings of Denmark and the Danish nobility related to one another in a complicated system of reciprocity. In accordance with an
unspoken, unwritten, but nevertheless well understood and seriously regarded contractual arrangement, the scion of a noble family
owed the king his loyalty and allegiance and a portion of his time
and talents. The king's obligation was to grant royal fiefs that provided substantial incomes. These grants were, in effect, the main
glue that bound king and aristocracy together. Weighing in heavily
on the side of the scales for a Brahe, Bille, and Oxe family member,
in this arrangement, were his potential personal contributions to the
welfare and prestige of the crown, the stature and power of the present extended family, and the valuable service rendered by previous
generations of Brahes, Billes, and Oxes. Tycho's uncle Jørgen Brahe
had lost his life as a result of saving this very Frederick from drown-

ing. Such acts were not forgotten when it came to the delicate question of who owed what to whom.

Tycho, though his abilities and interests lay in different directions from most of his aristocratic contemporaries, was as highborn and courtly in his manners as any young nobleman of his generation. King Frederick was not without appreciation of the fact that although Tycho had wandered from the traditional path, he had, by doing so, developed some unique talents to offer the crown. Frederick was a king who valued scholarly pursuits. One of his avowed goals was to make Denmark famous for its learning, and he had generously endowed the University of Copenhagen to make it one of the finest educational centers in the world.

The question now troubling the king and his advisors was whether Tycho, in addition to the promised canonry, should have a major fief. Was he equipped to manage one? Though he wore a sword and had fought a duel, and bore the scar from it, he was no warrior. He was not trained to ride into battle in armor or to command the defense of a castle, as his brothers were. Nevertheless, he clearly had to have something at least the *equivalent* of a fief. His powerful relatives at court were conferring with King Frederick and encouraging him to come up with an appropriate answer.

By the end of the year his father died, Tycho was centering his life not at court in Copenhagen nor at Knutstorp but at Herrevad Abbey, where his uncle Steen Bille lived. Herrevad, a few miles from Knutstorp across rolling countryside, meadows, and woodlands, was situated where the more settled area nearer the Øresund ended, and its lands bordered on the deep, limitless forest of the north. It had been a Cistercian abbey, founded in 1144, and a few aging monks were still there in Tycho's day. The abbey church probably remained an austere Romanesque structure on the outside, but the interior had been completely altered in the thirteenth century in a lovely Gothic style. The Billes, Tycho's mother's family, had several holdings in the area, but Tycho's uncle Steen had acquired Herrevad rather recently, in 1565. From its grand portal-house, suitably altered to be a noble dwelling, he administered the substantial estate that had once belonged to the Cistercians—hundreds of farms, mills, fisheries, and oak and beech forests where only the king was allowed to hunt the deer and stags. In addition to Steen Bille's contingent of knights and their squires, his wife, Lady Kirstine, and her handmaidens and servants, and all the support staff that went into maintaining such an establishment, there was also a Lutheran Latin school for boys on the premises.*

Steen Bille was a lively, outgoing man and extremely influential at the Danish court. He was also kind and gentle and associated happily with people of all social classes, especially relishing the company of scholars. With his encouragement and promise of active assistance, Tycho prepared to make good on his decision to take charge of his own career and research path. Uncle and nephew collaborated to transform Herrevad into a superbly equipped haven for the study of the subjects that interested Tycho and intrigued Steen as well. Setting up an independent research facility on this scale was unprecedented among Danish aristocrats, even among Danish scholars. Tycho and Steen Bille were treading new ground, and learning by trial and error.

Not only was there to be an astronomical observatory, but they also planned a paper mill, an instrument factory, and an alchemy laboratory. Alchemy, though it would eventually come to have disreputable connotations as a strange obsession with turning base metal into gold, in Tycho's day included medical alchemy and other related experimental science. It was the ancestor of modern chemistry. Tycho's interest was primarily in medical alchemy. Herrevad already boasted a medical curiosity, a rib bone six feet long, said to have come from the burial site of a man named Vene. "Verily," one scholar

---

*Herrevad today, still deep in the country, retains some of its medieval atmosphere, though it is now a riding school with only a small museum to recall the past. There are ancient trees, broken walls, and footings of vanished buildings, and shaded, pollen-strewn ponds that probably date from Steen Bille's day.

had commented, either seriously or wryly, "there did once live in these northern realms a people of wondrous dimension."

Alchemy required glassware, and a master of the Venetian art of glassmaking, Antonio de Castello Veneziano, with a retinue of assistants, arrived in Denmark, possibly on the run, for the Venetian Republic was extremely possessive of its glass industry. Steen invited them to make their home at Herrevad, adding Venetian dialect to the mix of Danish and Latin already spoken there. Soon the glassmakers were producing not only alchemy vessels for Steen and Tycho but also drinking glasses and windowpanes for King Frederick and Queen Sophie.

One of Tycho's first undertakings at Herrevad was to construct a new astronomical instrument, a "half-sextant" with straight walnut legs and a curved brass arc. A little later he added a larger, interchangeable sixty-degree arc. It was this sixty-degree arc that gave a "sextant" its name—probably coined by Tycho himself (see figure 3.1). Sixty degrees is one-sixth of a circle; a half-sextant has a thirty-degree arc. Sextants and half-sextants resemble slices of pie. By sighting along the two legs or sides (where the pie is "cut")—pointing one leg toward one star and the second leg toward another, for example—it was possible to measure the angular distance between two heavenly bodies. One could similarly measure a body's altitude above the horizon.

Tycho was getting more from this effort than a better instrument. He was developing expertise and learning lessons that would serve him later. One conclusion Tycho reached was that in order to design and manufacture instruments capable of the precision he wanted, he would need highly skilled, specialist instrument builders working at his own facility, where he could supervise them. For the moment, he had to content himself with getting the best results he could with nonspecialist artisans under his supervision, while ordering more intricate or decorative parts, and sometimes whole instruments, from Copenhagen. What began at Herrevad was never completely realized there, but the possibilities Tycho saw unfolding at the beautiful old abbey gave him a much clearer vision of what he hoped to accomplish and how to go about it.

In that same watershed year, 1572, when Tycho ended his European wanderings and began to pursue his interests closer to home, he met—perhaps not for the first time—a young woman named Kirsten Jørgensdatter. Like other key events in Tycho's personal life, the beginning of his lifelong relationship with her is frustratingly undocumented. She was not a woman of noble lineage. (If she had been, more would be known about her.) Pierre Gassendi, Tycho's earliest biographer, reported a description that he got firsthand from one of Tycho's last students: Kirsten was "a woman of the people from Knutstorp's village." To this day, tradition in the Knutstorp area has it that she was a clergyman's daughter and that her father was pastor of the Knutstorp parish church at Kågeröd, about two miles from Knutstorps Borg.

The name Jørgensdatter indicates that Kirsten's father's first name was Jørgen. From 1546 to 1569 the pastor at Kågeröd was Jørgen Hansen. Most likely it was he who christened the infant Tycho and buried his twin brother. If Kirsten was his daughter, she must have spent her childhood in the half-timbered parsonage beside the little stone Kågeröd church while Tycho was growing up in the castles of his uncle and father.

The family coats of arms of the Brahes and the Billes are carved on the arch of the Kågeröd church above the family pew. The pulpit in the church is situated not at the front but about halfway back, and the enclosed box pews have seats front and back so that if they are not too crowded, the occupants can move to face the pastor during his sermon. If Tycho sat in the family pew during his visits to Knutstorps Borg, perhaps his eyes fell on Kirsten as the congregation shifted their seats. The young daughter of the pastor would have been modestly clad, neither like a peasant girl nor like a child of the aristocracy. As she grew to be a young woman, a white lace collar and cuffs probably would have been her only less-than-somber adornments.

Much later in the lives of Tycho and Kirsten, after two other pastors had come and gone, a Hans Jørgensen—again indicating that his father's first name was Jørgen—was called to the Kågeröd church by the joint lords of Knutstorp, the brothers Tycho and Steen Brahe. Records show that this Hans Jørgensen visited Tycho at his island castle observatory that same year, 1591. Scholars have speculated whether he went there only to be interviewed for the position or also to visit his sister, who by that time had been living for twenty years as Tycho's wife.

Adding strength to the local tradition that Kirsten was the pastor's daughter, not a peasant girl, is the fact that though casual liaisons between noblemen and peasant girls were not unusual, lifetime alliances were. Tycho's relationship with Kirsten Jørgensdatter was no mere youthful dalliance. It is much more likely that he would have chosen as his companion for life the daughter of an educated clergyman, with a family background and upbringing not quite so drastically different from Tycho's own as a peasant woman's would have been.

If Kirsten was Jørgen Hansen's daughter, her position in society was indeed considerably above that of a peasant, but there was still a daunting chasm between her station in life and Tycho's. The nearest he had ever come to experiencing her world was when he lived with a clergyman during his school days, and that clergyman had probably been a bishop, certainly not a humble pastor. Kirsten would have grown up in obscurity in a thatched-roof cottage, working with her hands in the kitchen, house, and garden, and probably never traveling more than a few miles from the village of Kågeröd. So far apart were their worlds that Inger Oxe and Beate Bille would have had difficulty imagining what Kirsten's daily life was like, as Kirsten would have had difficulty imagining theirs.

However well Tycho and Kirsten would manage to bridge the gulf between their different upbringings, that gulf, as it was formally imposed by Danish society, law, and tradition, could not be bridged. Tycho the nobleman and Kirsten the commoner could not legally become man and wife.

There was an alternative that was considered neither scandalous nor sinful. The earliest Danish law codes of ancient Jutland had recognized the legality of *slegfred* marriages; that is, common-law or morganatic marriages. Under the ancient laws, which were still in force, a woman who was a commoner, who lived openly as a wife in a nobleman's house for three winters, dining, drinking, and sleeping with him and carrying the keys to his house, was his wife. Originally, among the polygamous Vikings, *slegfred* had meant a wife of secondary status, but in Tycho's day it had lost that connotation. When Tycho and Kirsten began their relationship, the courts had just recently reaffirmed that the offspring of such a marriage were not bastards but *slegfred* children. However, they and their mother remained commoners, no matter how nobly born the father was, and the children could not inherit their father's estates. None of the expectations and rights of a nobleman's sons and daughters applied to them.

Tycho was well aware that his choice of Kirsten strongly reinforced his image as a young man who was willing to flout convention, and that it would likely have drastic consequences for his future and for his descendants. The reputation and influence of his powerful extended family was also at stake. Family honor and alliances through marriage were of enormous importance to Tycho's relatives. Their reactions to his choice were, predictably, not enthusiastic. The only advantage for them was that the family inheritance would not have to be so widely shared. A few were actually sympathetic.

Surprisingly, Tycho's morganatic alliance with Kirsten did not dim his hopes at court. King Frederick had reason to be understanding. He, like Tycho, had fallen in love with a woman beneath his station, Anne Hardenberg, a noblewoman but not of royal blood. Frederick's father, King Christian III, had forbidden their marriage. After the old king's death, Anne had continued to live within the royal family as a part of the queen mother's court, and King Frederick had refused to have anything to do with negotiations for a different bride.

However, when Frederick finally announced that he would enter into a morganatic marriage with Anne, even though their children could never succeed to the throne, the opposition among the nobility at home and abroad was so vigorous that he in the end agreed to give up Anne and marry his fourteen-year-old cousin, Princess Sophie of Mecklenburg. Their marriage took place in July 1572, during the time when Tycho was falling in love and sealing his relationship with Kirsten Jørgensdatter. Frederick invited all the Danish nobility, commanding them to dress in new court attire and ride their best horses, and each to accompany himself with two squires and a page. Tycho—not being a warrior knight—had no squires and pages of his own and on such occasions had to borrow them from someone else in the family. Because Kirsten was not a member of the nobility, it was unthinkable that she could accompany Tycho to a royal celebration.

Tycho's earlier career choices had been unorthodox and had led him into astronomy. Now, unwittingly and many years before anyone could predict how it would all end, he had made a decision that would have repercussions far beyond his own lifetime, his own children, and the borders of Denmark. When Tycho took Kirsten Jørgensdatter as his life partner and began to sire children by her, he set his feet much more firmly on the path that would lead to Prague and to Johannes Kepler.

A plaque on an exterior wall at Herrevad commemorates an astronomical event late that same year that also gave Tycho a powerful push toward that future. The plaque announces: HERE TYCHO BRAHE, ON THE EVENING OF NOVEMBER 11, 1572, DISCOVERED A "NEW STAR."

It was a clear autumn evening after several days of overcast skies. Tycho, now nearly twenty-six, was walking back to supper from his alchemy laboratory, glancing up at the familiar darkening sky as he went. To his astonishment, right over his head, near the three stars that make up the right-hand half of the *W* of the constellation Cassiopeia, there was a star he had never seen before. "I knew perfectly well—for from my youth I have known all the stars in the sky, something which one can learn without difficulty—that no star had ever before existed in that place in the heavens," Tycho wrote, "not even the very tiniest, to say nothing of a star of such striking clarity." It was brighter than any other star or planet in the sky.

Not quite trusting his eyes, and wanting witnesses to what he was seeing, Tycho called his servants and then stopped some peasants who were passing nearby. These people had not spent nights studying the stars as Tycho had, but they dutifully craned their necks to gaze up beyond the trees and the darkening walls, trying to oblige their noble companion by giving him an opinion as to whether or not this really was something new. They were not able to confirm that this star had not been there before, but they did agree, when Tycho called their attention to Venus, that the new star was brighter even than that bright planet. "I doubted no longer," reported Tycho. "In truth, it was the greatest wonder that has ever shown itself in the whole of nature since the beginning of the world, or in any case as great as [when the] Sun was stopped by Joshua's prayers."

Tycho realized that the star's position in relation to the zodiac meant it could not be a planet, and, though he had never seen a comet, he knew from his reading that a comet has a tail and a fuzzy appearance. This had neither. However, the real test of whether it was a comet was whether it moved in relation to the other stars. Finding whether it did took several nights of watching, armed with his cross staff. Tycho could not discern any change of position. This was no comet. Though more observation and calculation were needed to make certain, Tycho was also fairly confident that the new star was not closer to Earth than the Moon's orbit—a dramatic conclusion in the context of the astronomy he knew. "Let all philosophers, new as well as ancient, be silent! Let the very theologians, interpreters of the divine mysteries, be silent! Let the mathematicians, describers of the heavenly bodies, be silent!" he exclaimed.

Aristotle's ancient cosmology, which insisted that change could occur only in the region closer to Earth than the Moon's orbit (the "sublunar" region), was still gospel among most scholars. Tycho had never declared himself an avid follower of Aristotelian cosmology, but he had not escaped its influence. However, it was in Tycho's nature to want to test things out for himself. To do that, he had to try to measure the parallax shift of the new star, or *nova,* as he dubbed it.

Parallax shift is the apparent shift of an object against the background when observed from different viewing positions. The simplest demonstration is by holding one finger up in front of your face, focusing on the distance, and closing first one eye and then the other. The finger appears to shift from side to side against the background. Your two eyes are the two "viewing positions." The shift is a parallax shift. The further away you hold your finger from your face, the smaller the shift.

Though scholars had understood for millennia the mathematical principles of such a shift and of the way it grows smaller with distance, determining the distance to a star by parallax was still impossible in Tycho's day. Powerful enough telescopes would not exist until the 1830s. However, astronomers had known since ancient times that the Moon does have a parallax shift, that two observers a distance apart on the face of the Earth see the Moon in two different positions against the background stars. An observer could even stay in the same place and allow the daily rotation of the celestial sphere (or the rotation of the Earth, if he believed Copernicus) to change his "viewing position" for him. If he did that (as Tycho had done), he saw for himself that the Moon did indeed have a parallax shift. A star or other object as close as the Moon, or closer, would also have a parallax shift. And so, although it was not possible for Tycho to find the distance to the new star by means of a parallax measurement, it *was* possible for him to conclude that if this new object in the sky did not show any parallax shift against background stars, while the Moon *did*, it could not possibly be nearer than the Moon.

Tycho was not the only person who noticed the nova, and not all agreed that it was further away than the Moon's orbit. Some actually made observations and were convinced that the new star was below the Moon, even though they could discover no parallax. Some insisted it was a comet. Others conceded that it was farther away than the Moon but argued that it was not really new, or that it was not a real change in brightness that caused it gradually to dim after its first appearance.

Evidently, Tycho was the only scholar capable of seeing beyond Aristotelian assumptions. He measured the angular distance between the nova and another star, and repeated the measurement several hours later. Though the sky as a whole had moved, the distance between the two stars had not changed. Tycho performed this test more than once, measuring the angular distance between the nova and not one but several other stars. Again and again he found no change, no parallax shift. Because this result ran so counter to current science and philosophy, Tycho decided to come at the problem from another direction. He had already calculated, indirectly—by comparison with other stars—how far the nova was above the celestial equator in terms of angular distance (its declination). Now, as a double check, he measured the declination directly, which required finding the maximum height the star went above the horizon at Herrevad. If the nova had never gone higher in the sky than sixty degrees above the horizon at Herrevad, he could have used his sextant, which measured sixty degrees. However, the star went much higher. It reached its highest point at Herrevad only six degrees from the zenith (the point of sky directly above Tycho's head).

Tycho's way of solving this problem, obvious as it seems in retrospect, was a true innovation. He turned the sextant around, set it in a north window, recorded the star's lower culmination—that is, how far above the horizon it was at its *lowest* point—and calculated the star's declination from that. He was so pleased with this simple ingenuity that he included a drawing of the sextant in that position much later in his book celebrating his instruments (figure 3.1).

Before a month had passed, Tycho had completed these tests to his own satisfaction and was confident enough to send his findings and his conclusions—that the star was not a comet and not in the sublunar region—to a few friends. He was not so sure of himself as to send them to scholars in the more learned circles at the University of Copenhagen.

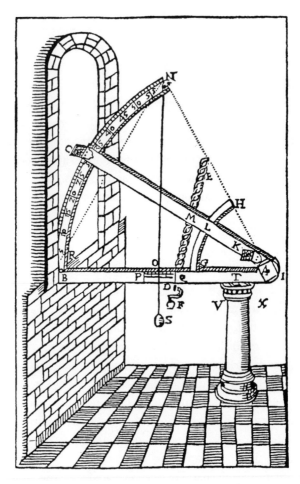

Figure 3.1: One of the first instruments Tycho designed was the sextant, used for measuring altitudes of bodies above the horizon, their azimuth (distance from the meridian), or their distances from one another. Two legs were joined with a hinge (right) so that the end of one moved along the arc (left) by means of a screw (*E*). There were sights at *C* and *K*. This drawing from *Astronomiae Instauratae Mechanica* (probably of Tycho's second sextant) shows it set in the north-facing window to observe the nova at its lower culmination. He put the hinge end (*I*) close to his eye and found the star through the two sights.

As his twenty-sixth birthday approached, rather than spend more time and effort reporting his study of the star, Tycho assigned himself an entirely different project that had to be completed before the end of the year or not at all: putting together an astrological meteorological almanac for 1573. If the predictions in it were to have any impact, it was necessary to complete it before that year began.

Tycho's introduction to his almanac was in the form of an "oration" that provides a window into the mind of this eloquent, thoughtful, and well-educated young man. He began it in traditional fashion by invoking Urania, astronomy's muse in classical mythology, and proceeded to describe the universe as being Earth, seas, Sun, Moon, stars, animals, vegetables, and minerals, and the Creator as incorporeal, immense, eternal, incomprehensible, and omnipresent but not located in any single place. It was a good Lutheran beginning.

Tycho voiced the conviction that humans were created in the image of God, by God, who put them on Earth at the center of the universe with a good view of the rest of it, so that through contemplation of the visible Creation they might learn something of the majesty and wisdom of the Creator. There was, Tycho declared, no better teacher of theology than the universe itself, and this was so even on Earth, where, with the exception of the human soul, dissolution and change held sway. It was especially true in the celestial realms—immense, unchanging—so redolent of God's power and intellect. In view of all this potential, Tycho lamented the ignorance of most people—even those who fancied themselves as authorities—about the heavens.

He continued by discussing his theory of meteorology, the theory that would underlie his almanac: celestial events, especially of the Moon—because it is immediately next to Earth's atmosphere—had, he thought, a strong influence on the weather. However, one should not put too much reliance on weather prediction based on celestial events because there were also local conditions on Earth, this realm of chance and change, and these conditions varied from place to place. Therefore, his hope was not so much to predict the weather as to study the *discrepancies* between predicted and actual weather and find out more about how Earth and heaven were linked. Thus Tycho neatly sidestepped the danger that his predictions might not be correct. Discrepancies weren't a problem; they were the leading edge.

Tycho also managed, in passing, to sneer at some of his contemporary astronomers and to pay homage to Danish royalty. He said he was not one to sit snug by the fire and learn his astronomy from books and papers, and so he had decided to use his own observations rather than the Alfonsine (Ptolemaic) or Prutenic (Copernican) tables. Furthermore, because all people owe a great debt to their native land, he would use Copenhagen as his place of reference in establishing the meridian and the horizon. In other words, for purposes of this almanac, King Frederick should think of himself as the center of the universe.

At the end of his introduction, Tycho listed the other manuscripts he had written on subjects related to the topic at hand. He had been busy. There were tables of the risings and settings of the Sun, Moon, and planets, their configurations for each day and each lunar octad, and the predicted weather for each day of the year. He urged the need for systematic meteorological observations, and then he moved on with a flourish to quote Ovid on the joys of astronomy. Finally there was a verse he had written himself, lamenting that the demands of the world, of courtly life, and the intense cold of the north disturbed the serenity that a man required for contemplating the stars.

As work on the almanac progressed, Tycho ran into a hitch. The almanac predicted an eclipse of the Moon on December 8, 1573. Investigating the astrological significance of this event, he found that it seemed to predict the death of King Frederick. Such a prediction was a matter of national security and certainly could not be announced straightforwardly in an almanac. On the other hand, if the king died, it would be a feather in an astrologer's cap to have predicted it. Tycho chose to clothe the announcement in rather garbled allegorical writing, which if necessary could later be interpreted, with hindsight and a little help from its author, to predict this catastrophe. (It did not occur.)

Soon after the new year began, Tycho carried the almanac manuscript to Copenhagen along with several other manuscripts, including the one about the new star. To his astonishment, he learned that no one in Copenhagen had yet spotted the nova. He spoke of it while dining among friends at the home of Charles de Dançey, the French ambassador. Dançey thought Tycho was chiding them all for not watching the sky as closely as he was. Johannes Pratensis, who was on the faculty of medicine, sniffed that other professors at the university could not have missed anything so dramatic. Tycho held his peace and allowed the conversation to move to other matters. When a clear night came, there was the star, and it was, as Tycho had told them, not at all like a comet. Amid the excitement, Tycho brought out the manuscript he had written about the star. Pratensis urged him to publish it.

Tycho had never published any of his manuscripts, nor was he thinking of publishing this one. Scholarly endeavors were beneath a man of his rank. A nobleman might read a book and acquire some learning thereby, but surely not write one. And if he did happen to write one, it was surely for the sport of it, not for public consumption. Tycho brushed off Pratensis's suggestion on the grounds that his manuscript was not polished enough, that he had never intended it to be widely read.

Pratensis did not give up. After Tycho had gone home to Herrevad, Pratensis sent him some reports written about the new star that had only recently reached Copenhagen from abroad with the spring thawing of the sea-lanes. When Tycho read them, claiming he did so only because he lacked anything to do while he was ill, he was distressed by the incompetence of those who had studied the star, and particularly with claims that it was a comet only about as far away as twelve to fifteen times Earth's radius.

Back in Copenhagen to check on work being done for him in an instrument shop, Tycho again encountered Pratensis. Only now did he admit to Pratensis (who was not a nobleman) what the real obstacle was. Pratensis didn't take offense, and suggested Tycho consult his powerful relative Peder Oxe. Oxe raised the possibility that Tycho might publish his manuscript anonymously, but Tycho returned home again with it still under his arm. In the end, Pratensis prevailed, and a letter from him became the preface to the published book.

Tycho's decision to publish *De Stella Nova* was a major turning point. To his mind and the minds of his contemporaries, a life of serious scholarly pursuit and the life of a nobleman were incompatible. If he had got by so far, it was because he still seemed a young man who had not yet settled down seriously to either sort of life. However, it had in fact become far too late to follow the traditional path into knighthood, and government service—which was still not out of the question—was simply not what he wanted. Hence he was already some distance beyond the pale even before he decided to publish this book. Nevertheless its publication was, for him, another significant and overt step away from orthodoxy.

De Stella Nova began as a quiet, short treatise, not passionate or provocative. Tycho reported his studies and his findings and reasoned about the location and the nature of the star. He also dealt with the purpose of it, a matter that greatly interested him—not merely its philosophical meaning but also its practical significance— for surely such a celestial event could not be completely irrelevant to humans. The astrological predictions turned out to be both difficult and dire, but more telling when it came to Tycho's own future was a last addition to the manuscript before it went to press in the spring of 1573, when the star was still visible but greatly dimmed.

Though the front of the book included a poem by one Professor Johannes Franciscus praising Tycho's noble lineage, Tycho chose in his epilogue to cast scorn on the supposed glories of his class: feats of arms, association with kings, and the pursuit of frivolity, wine, woman, and song. He declared himself unwilling to be held back by fear of what people would think of him, for his would be the greater and immortal glory of having improved astronomy beyond anything it had been before. His lineage was among the noblest, but he himself would take pride only in what he would accomplish himself.

Tycho concluded with an allegorical flourish. He portrayed the goddess Urania commissioning him to find the position, distance, and meaning of the new star, but that was only the beginning. He also was charged to do the same for all the other stars, plot the paths of the Sun, Moon, and planets, and discover their influences on meteorological phenomena. Who, he asked, having such a vision, could ever lower his eyes to mundane interests? Tycho sent copies of the book to friends, patrons, and scholars, but none to his family. Many years later, Johannes Kepler wrote that if it signified nothing else, the nova of 1572 heralded the birth of an astronomer, the great Tycho Brahe.

☙

I N   1 9 4 5 , astronomer Walter Baade made it a project to find out what the bright object was that Tycho discovered in the sky over Herrevad in 1572. Studying Tycho's observations, Baade concluded that it was in all probability a Type I supernova, the explosion of a "white dwarf" star. Such a cataclysm occurs when an elderly star has exhausted all its nuclear fuel and collapsed to a sphere about the size of Earth, with a mass close to the mass of the Sun. In a star that small with a mass that huge, matter is packed to almost inconceivable density—hundreds of tons per cubic inch. Most white dwarf stars are parts of "binary systems" in which two stars continually circle one another. The dwarf's partner in the system is usually a much larger but far less massive star. As the two perform their celestial waltz, the denser dwarf star cannibalizes matter from its companion and gradually adds to its own mass. The mass limit a white dwarf can attain without collapsing and exploding under the pull of its own gravity is about 1.44 times the mass of the Sun. Once having exceeded that limit, the star rips apart in a titanic explosion. That explosion is a Type I supernova.

Tycho knew nothing about supernovae. Knowledge about them was more than 350 years in the future. Tycho called the star *nova*, Latin for "new," though modern astronomy uses the term *nova* for a less violent explosion.

Type I supernovae happen frequently in the universe, but they are rare in any one galaxy. In our Milky Way there was one in 1006. There is no record of anyone in Europe seeing it, but in China it was called a "guest star." Tycho's nova was next in 1572, and in 1604 there was another, observed by Johannes Kepler. There have been no Type I supernovae in the Galaxy since. Tycho knew of no precedent except for a report of a new star in the second century B.C. from the Hellenistic astronomer Hipparchus of Nicaea. Tycho ruled out the possibility that the star that heralded Christ's birth and led the Magi was another example, on the grounds that that star had to be much nearer to Earth to lead anyone anywhere. He was again underlining the fact that the 1572 nova did not move in relation to other stars, and hence was extremely distant, well beyond the Moon.

Radio astronomers in the late twentieth century were able to identify a source of radio emission that they believe comes from what remains of the 1572 supernova. They place it exceedingly far beyond the orbit of the Moon.

4

HAVING THE BEST OF SEVERAL UNIVERSES

1573–1576

SPRING, WITH THE THAWING of the sea-lanes, was the customary season for a Dane to set off on travels abroad. In spite of a new wife and the work at Herrevad, Tycho was restless again in the spring of 1573. He made plans to leave Denmark and was even thinking of making a permanent move. Details of his book's publication delayed the journey . He still had not left by the following autumn, when Kirsten gave birth to a baby girl, also named Kirsten, on October 10.

On November 11, Pratensis held a Martinmas feast at his lodgings in Copenhagen to mark the first anniversary of the evening when Tycho first saw the nova. One copy of the invitation still survives, promising sugar, almonds, chestnuts, a goose, a suckling pig, and eighty bottles of wine. They were going to drink more than they were going to eat. Although Tycho may have poured scorn on the frivolities and excesses of the Danish nobility, life among scholars was not exactly abstemious.

Tycho did little observing in the summer and autumn of 1573, partly because of the distractions of publishing his manuscript. However, in early December he and his fourteen-year-old sister

Sophie observed an eclipse of the Moon and discovered that some adjustments he had made to the Prutenic (Copernican) Tables, while calculating when the eclipse would occur, had succeeded even better than he had hoped. "I myself cannot sufficiently marvel over the fact that at this early age, only twenty-six, and without the aid of numerous and accurate observations of the motions of the Sun and Moon, I should have been able to obtain such precise results," wrote Tycho with unembarrassed self-approval.

Tycho and Sophie observed the eclipse from Herrevad with an elegant new quadrant built to his order by specialists in Copenhagen. It was a creation of great artistic beauty, fashioned of brass and gold. Evidently Tycho's financial situation had improved now that he could anticipate inheriting, soon, a portion of his father's estate.

That inheritance notwithstanding, a painting on the side of the quadrant reemphasized that Tycho had little good to say of his noble heritage. It showed a table holding symbols of aristocratic life—scepters, coats of arms, ostentatious clothing, goblets, dice. Around the table were symbols indicating the futility of that life—a skeleton, a withered tree. On the other side of the tree the branches and roots were shown alive and abundant, and seated in its shade was a man studying a book and a celestial globe. "By Spirit we live," declared Tycho's inscription above the picture. "The rest belongs to death."

The new quadrant wasn't large or designed to make such fine measurements as Tycho's sextant, and it must not have been very useful, for he made few observations with it. However, Tycho clearly felt that instruments made of metals rather than wood represented the future. The quadrant was a first experiment with a relatively inexpensive model and also an exquisite trophy to celebrate his new life as an astronomer and his break with the traditions of nobility.

During the summer of 1574 Tycho moved to Copenhagen. There is no record of Kirsten and his daughter accompanying him, nor any to show whether Kirsten was "keeping the keys" of another house in accordance with Jutish law. Knutstorp was Tycho's widowed mother's

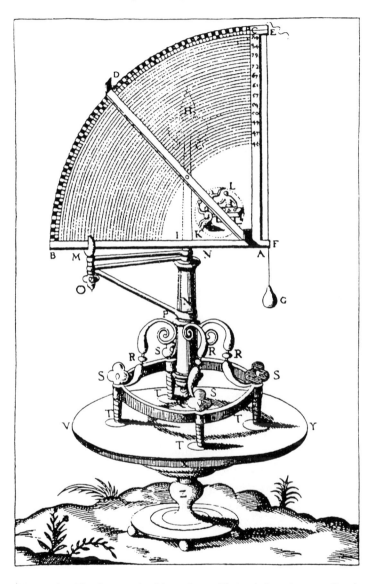

Figure 4.1: The brass-and-gold quadrant. The painting that repudiated Tycho's aristocratic heritage is in the circle marked *K* and *L*. The drawing is from *Astronomiae Instauratae Mechanica*.

domain, Herrevad his Uncle Steen's and Aunt Kirstine's. But Kirsten was expecting their second child. The baby girl was born before the end of the year and christened Magdalene.

Tycho's status within the Copenhagen scholarly community had been transformed with the publication of his book. *De Stella Nova* was a serious professional credential and established him as something of an authority in astronomy. Before the summer ended, there was talk of his delivering lectures at the university, even though he had never acquired an M.A. degree or intended to have an academic career.

There had been obstacles to a nobleman publishing a book. There were even greater obstacles to his lecturing at the university. Not only his own image and self-image were at stake but the status of others as well. Since the Reformation, church and university had become the domains of an educated middle class—usually better educated than the nobility—who served as bishops, lesser clergy, teachers, professors, and scholars. During Tycho's lifetime and for many years thereafter no nobleman in Denmark served in these capacities. An unwritten rule of society in sixteenth century Europe was that people, knowing their places, stayed in them and did not trespass on other territory. It was almost as unthinkable for a nobleman to move down the ladder as for a commoner to move up. It was not exactly forbidden; it just did not happen.

Someone thought of a way around this obstacle: A number of noble university students signed a petition and presented it to the king, who then added his request to theirs, inviting Tycho to lecture to these students of his own social class. The lectures were also open to anyone else who chose to attend. After that matter was settled, it did not take long to schedule the lectures.

The first, held on September 23, was an hour-long formal introduction to the series. Two years earlier, Tycho had revealed his philosophy and beliefs in the introduction and epilogue of his almanac. Now, in 1574, approaching age twenty-eight, he was still a religious man, deeply influenced by the religious traditions of his time, continuing to interpret human activity within a context of religious belief. He spoke of Seth and Moses before moving on to Hipparchus, Ptolemy, and Copernicus. As he had done in his almanac, he argued for the value of astronomy in liberating human minds from worldly matters and turning them toward heaven, as well as for providing calendars and predicting the weather. That last usefulness had been called into question because of so many failed predictions. However, Tycho pointed out that the Sun was responsible for the seasons of the year, and many believed (though not everyone in his time— Galileo was to be a notable exception) that the Moon influenced the tides. To Tycho it seemed reasonable to think that the stars also had to have something to do with the turbulence of weather and other weather patterns.

Discussing the use of astronomy for horoscopes was a touchier matter. Since Augustine of Hippo in the fourth and fifth centuries, nearly all Christian theologians, including Luther, had opposed the use of astrology to predict human events. Tycho felt obliged to rebut this long tradition and to come down on the side of Luther's disciple Melanchthon. When Tycho began this part of the lecture, nearly all in the room turned their heads to look at Niels Hemmingsen. That elderly theologian and follower (usually) of Melanchthon had recently switched his allegiance and attacked Melanchthon's favorable views about astrology. Hemmingsen smiled and tipped his academic hat. Then the attention reverted to Tycho.

Tycho took this hat tipping as a delightful challenge. Philippist theology (the theology of Philipp Melanchthon) had a strong hold at the university. Tycho hoped to show in what way his own views meshed with this popular thinking and how they overcame Hemmingsen's objections to it. The heart of Tycho's argument was that the stars influenced individual lives and human events but did not *determine* them. He insisted that "there is something in man that has been raised above all the stars." God had endowed human beings with free will, and created man so that he can "overcome any malevolent inclinations what-soever from the stars if he wills to do so." Tycho recommended good education, discipline, and other desirable human activities, for in them—in rising above a "brutish life"—lay the best possibility of de-flecting the influence of the stars. Tycho would never lose his faith in astrology, interpreted as he did in this lecture, even much later when he was focusing almost exclusively on astronomy.

Tycho was a complete novice when it came to talking before a crowd, but he must have been a riveting speaker, for after the lecture a professor of law, Albert Knoppert, approached him and commented, "When I heard your attacks on the philosophers and physicians, and even the theologians, I was afraid that you would also launch into us jurists—so afraid that I broke into a sweat!"

The lecture crowd dispersed, but the discussion about astrology and free will continued at a leisurely meal hosted by Dançey. Hemmingsen wanted to clarify a few points—particularly that Tycho agreed that God works and acts with absolute, unrestricted freedom and that hu-man beings also have completely free will—and then declared himself well satisfied with Tycho's views.

The day following this highly successful introduction, Tycho be-gan in earnest to lecture on astronomy. He kept no notes, and proba-bly used none, except to record that he covered the theories of the Sun and Moon "according to the models and parameters of Copernicus" and supplied copies of portions of the Prutenic Tables for those in the audience who could not afford to buy them.

Tycho had not, however, rejected Ptolemaic astronomy and be-come a convert to Copernicanism. He explained Copernicus's Sun-centered planetary theory, but then he proceeded to explore the pos-sibility that Copernican theory could be interpreted in such a way as to allow Earth to stand still. In Tycho's words, Copernican theory might be "adapted to the stability of the Earth." A decade would have to pass before he would realize this early hope in his "Tychonic system," but Galileo's nemesis was waiting in the wings.

Tycho's attitude toward Copernicus—deep respect that fell short of acceptance of a moving Earth—would not have seemed a startling departure to the more well informed among his audience. Although Copernicus's book *De Revolutionibus,* which had been published a month before that astronomer's death in the spring of 1543 (about thirty years before Tycho's lectures), departed radically from the prevailing Ptolemaic worldview, neither the scholarly nor the religious world had reacted negatively to it. Galileo's clash with the pope did not occur until well after Tycho's death. Meanwhile, a great deal of time was elapsing, and no overt, dramatic conflict was occurring among scholars or theologians about how the universe was arranged.

Copernicus himself undoubtedly believed that his Sun-centered astronomy represented literal truth—what really was going on in the universe. He did not regard it as only a mathematical scheme that made the prediction of planetary movement simpler and more accurate. However, it was not in the mind-set of his contemporaries to as-

Nicolaus Copernicus

sume or even to recognize that he had made any such truth claim in *De Revolutionibus.* Aristotle had done ancient and medieval astronomers a considerable service by drawing a line between physics and the mathematical sciences, including astronomy, in a way that could be interpreted to mean that astronomers need not search for Aristotelian "causes" for celestial motions. By Ptolemy's day, it had become routine to invent devices such as the epicycle and equant that yielded reliable predictions, without any need to explain what might cause the planets to move in the manner prescribed by those devices. In fact, to declare that Ptolemy either did or did not think the planets literally move in the way these mechanisms had them moving would be to misunderstand him. In the absence of any remote chance of conclusive direct evidence one way or the other, there was much to be said for not belaboring that question—maybe for not even realizing the possibility of such a question. A man who worried about whether his mathematical system represented literal reality was an exception. This was not an intellectual situation confined to the ancients. A similar mind-set exists today at the leading edge of theoretical physics.

Copernicus had no better view of the skies than Ptolemy, and his forte was not observational astronomy: Most of his observations were less accurate than those of Hellenistic and Islamic astronomers. However, Copernicus accepted the neo-Platonic idea that underlying all the complication of nature there are simple, harmonious patterns. Nothing so complexly convoluted as Ptolemaic astronomy and the mathematical wilderness needed to support it could represent truth. Copernicus's own system, which potentially explained planetary movement in a much more spare and simple way, must, he believed, therefore be a clearer insight into reality.

However much Copernicus was motivated by a desire for greater simplicity, *De Revolutionibus* was not an easy book, and not many had the expertise to wade through it, but skilled mathematicians soon found that Copernicus's math was brilliant and extremely useful. Long before controversy arose about whether or not Earth really moved and whether it or the Sun was in the center, Copernicus's math and astronomy had proved too valuable to discard. The Copernican Prutenic Tables, which Tycho had found a little more reliable than the antiquated Alfonsine Tables based on Ptolemaic astronomy, were drawn up in Protestant Wittenberg by Melanchthon's younger colleague, Erasmus Reinhold. Melanchthon and others of his school revered Copernicus for his mathematics and tables, and used these without worrying themselves about any deep cosmological conflict. Catholic scholars used Copernicus's tables to calculate a new calendar, and an amicable situation continued in which the Catholic Church had no official policy regarding the arrangement of the cosmos. It had managed to stay largely clear of that debate already for at least three centuries, for Copernicus had not been the first to propose a moving Earth. There was a vague, tacit understanding that if all parties could avoid declaring that any scientific arrangement of the universe or any scriptural cosmological statements should or should not be treated as literal truth, then everything would continue smoothly. That made a great deal of sense when there was no observational evidence to argue definitively for either theory.

The blame for Copernicus's failure to make clear his conviction that his theory represented reality—so that no one could have misinterpreted him—lies most directly with a preface to *De Revolutionibus* that was added by Andreas Osiander, who was left in charge of overseeing the final stages of publication. Osiander urged Copernicus to write a preface claiming that his theory was intended to be interpreted hypothetically. Copernicus refused, so Osiander wrote it himself anonymously, implying that Copernicus had penned the warning, "Beware if you expect truth from astronomy lest you leave this field a greater fool than when you entered."

If Copernicus saw the preface, he was nearly on his deathbed and may not have understood what it was. It undoubtedly would have angered him, but it meshed with the mind-set of his time and was instrumental in allowing his astronomy to infiltrate the scholarly world without seeming to pose a threat. Astronomers with whom Tycho had come into contact as a student drew on both theories without feeling they were engaging in intellectual contradiction. Tycho was able to study Copernican astronomy with awe and respect, while not agreeing that Earth moved and the Sun was the center of the system. His use of both Ptolemaic and Copernican tables was not unusual.

However, by the early 1570s, even before the nova led to any crisis of faith in Aristotelian/Ptolemaic cosmology, a few younger scholars had begun to explore more seriously some of the implications of Copernicus's theory. If Earth was in orbit around the Sun, then observers on Earth ought to see an annual shift in the apparent positions of the stars. This was not a new idea. Since ancient times the lack of this "stellar parallax" shift had been taken, very scientifically even by modern standards, to mean that Earth did not move. Similarly, if one were in a carriage, looking out at a forest, and saw no shift of the tree trunks in relation to one another, one could be fairly certain the carriage was not moving. There was another possible explanation for the absence of stellar parallax, one that the ancients had not been willing to accept—except for Aristarchus of Samos, who had originally proposed the idea of a Sun-centered cosmos seventeen centuries before Copernicus: Suppose the stars were infinitely or nearly infinitely far away. Riding in the same carriage, but with the forest extremely distant on the horizon, one would discern practically no shift of the tree trunks in relation to one another.

Logic therefore dictated that accepting Sun-centered astronomy as literal truth meant accepting also that the universe is enormously large. With this realization, new speculation began about the distances of the planets and the stars. The nova of 1572 contributed to that speculation as well as challenging belief in the unchangeable perfection of the celestial realms.

Copernicus had concerned himself only a little more than Ptolemy with questions about *why* the heavenly bodies should move as they do. *How* they move was puzzle enough to occupy a man for a lifetime, let alone the question of what caused them to move that way. However, it was not long before it began to seem a little absurd to suppose that the universe could be working according to both the Ptolemaic and the Copernican systems at once. In hope of preserving some semblance of logic, scholars began in earnest to explore whether it had to be one or the other and to ask whether there was a way to draw old and new ideas into a coherent picture. This was the question Tycho had decided to try to answer.

Tycho was one of those younger scholars who believed that astronomy had to be more than brilliant mathematical constructs, that its goal could and should be to reveal what is really happening. He also recognized that Copernicus thought he *had* revealed that. Nevertheless, Tycho felt obliged to reject Copernicus's idea of a moving Earth. Because Tycho regarded Copernicus as the greatest mathematician of the century, this rejection, rather than being a sign of closed-mindedness, was a symptom of his independence and self-confidence as a scholar. He was not being antiscientific or militantly ignorant. The only way to prove that Earth was not standing still was to find stellar parallax, and that Tycho could not do. There was simply no physical evidence to show that Copernicus was right.

Earlier, when he was a student in Leipzig, Tycho had come to agree with Copernicus that Ptolemy's use of the equant was offensive, and it seemed true that Copernican astronomy had removed the need for this theoretical point around which a planet appeared to move without changing its speed. So Tycho had begun to search for a way to eliminate the use of the equant without capitulating entirely to Copernican astronomy. Because he did not use notes for his Copenhagen lectures, it is only possible to surmise from later developments what he might have said. He had declared his intention to expound "the motions of the planets according to the models and parameters of Copernicus, but reducing everything to the stability of the Earth," thus avoiding both "the mathematical absurdity of Ptolemy and the physical absurdity of Copernicus." The result toward which he was moving at the time of the lectures was a mechanism resembling figure 4.2, although he may not yet have had this scheme firmly in mind.

❦

IN THE SPRING of 1575, when Tycho's father's estate was finally fully settled and Tycho came into an annual income of about 650 dalers, approximately twice the annual stipend of a senior professor at the University of Copenhagen, he abruptly canceled the remainder of his lectures. Once again he had decided to go abroad, this time in style, accompanied by servants and a baggage train. Kirsten stayed at home in Denmark.

Tycho went with the blessing of King Frederick and, in a sense, as Frederick's emissary. The king had grandiose plans to transform the dark medieval fortress that guarded the entrance to the Øresund at

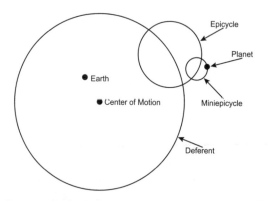

Figure 4.2: One of Tycho's first attempts to explain planetary motion without using an equant but leaving Earth as the unmoving center of the system: He moved the center of motion to a point some distance from Earth. He also added a miniepicycle. The planet rides on the miniepicycle, which wheels along centered on the epicycle, which in turn wheels along centered on the deferent—which is centered on a point some distance from Earth.

Wilhelm IV, Landgrave of Hesse, in a 1577 painting, probably by Caspar van der Borcht. In the lower right-hand corner there is a small portrait of Tycho Brahe, shown holding an instrument that is probably his sextant.

Elsinore (immortalized in Shakespeare's *Hamlet*) into a Renaissance castle. Finding Europe's best architects, hydraulics engineers, decorators, smiths, sculptors, painters, weavers, and myriad other experts required some scouting, and this was a task for which Tycho was well suited. Tycho's personal reason for going was quite different and would not have pleased Frederick. He was looking for a place to put down roots and pursue his own scholarly passions on a more extensive scale than he could at Herrevad.

Tycho stopped first that spring in Kassel with Wilhelm IV, Landgrave of Hesse, who was an avid astronomer himself. Wilhelm had commissioned many astronomical instruments, clockworks, and gadgets, and he conducted something like an academy for the advancement of astronomy. Overburdened with affairs of state, he had recently found little opportunity to make astronomical observations, but when Tycho arrived he set aside all other activities. Wilhelm agreed with Tycho that the nova had been beyond the Moon, though he thought he *had* detected a bit of parallax. He commemorated Tycho's visit by having a small picture of Tycho painted in the background in his next official portrait.

From Kassel, Tycho continued to Frankfurt to add to his personal library at the famous Frankfurt Book Fair, a long-lived trade fair that was a staple of the medieval intellectual world and still occurs annually. Venturing farther south, he revisited Basel. Peder Oxe, Charles Dançey, Steen Bille, and Johannes Pratensis all had studied at the University of Basel. Tycho was favorably impressed with the mild climate and easy access to scholars in France, Germany, and Italy.

From Basel Tycho traveled south once again and eventually reached Venice, where more than thirty years later Galileo would astound the Venetian doge and senate with a new invention, a telescope. Tycho was invited to learned gatherings that were part of the lives of rulers of the Venetian republic and which also included scholars from the university in nearby Padua. As both a nobleman and a scholar, he was in his element. In the interest of discovering experts for King Frederick, Tycho probably viewed in person some of the new villas designed by Andrea Palladio in the Veneto, the mainland near Venice. These architectural gems, whether he saw pictures of them or went there himself, made a lasting impression on him.

When summer ended and before the snows closed the passes through the Alps, Tycho journeyed north again to Innsbruck and then to Augsburg, where he had previously spent such enjoyable and profitable months. At Regensburg, not far downriver from Augsburg, he attended the coronation of Rudolph of Hapsburg, a ruler destined to play a significant role in his future. Rudolph was being crowned "King of the Romans," which made him heir apparent to the Holy Roman Empire.

Having considered numerous possibilities during his journey,

Tycho decided he would emigrate from Denmark and settle in Basel. On the way home to put his plan in motion, Tycho's final stop on the Continent was in Wittenberg. There he found university and city astir with news of a religious conflict, an early eruption of a division among Protestants that would later cause great unhappiness for Johannes Kepler and make working for Tycho Brahe his only option.

Twenty years earlier, the 1555 Peace of Augsburg had ended a period of widespread religious and political upheaval—or at least much of Europe had hoped it would. According to the treaty, only Catholicism and Lutheranism would be tolerated within the Holy Roman Empire. The present trouble had begun when Augustus, elector of Saxony, discovered that many theologians at Wittenberg were secret followers of another Protestant reformer, John Calvin. Augustus took strong measures. He imprisoned the errant theologians, including Melanchthon's son-in-law. When Tycho reached Wittenberg, they were still in prison, and friction had developed between Augustus and King Frederick of Denmark.

Denmark had not been party to the religious articles of the Peace of Augsburg, and Niels Hemmingsen, the elder scholar who had tipped his hat at Tycho's lecture, had been instrumental in achieving a consensus in Denmark that made no sharp distinction between Lutheranism and Calvinism. Hemmingsen also had published a work that gave a Calvinist interpretation to the Eucharist, and a similar work had appeared anonymously in Wittenberg. Worse for Denmark, all the imprisoned theologians in Wittenberg admitted that they had obtained their Calvinist ideas from Hemmingsen while attending the festivities surrounding King Frederick's wedding in 1572. Hemmingsen appears to have been an exceedingly busy and effective proselytizer who did not pause for royal holidays.

Since the trail of clues led directly to Copenhagen, Elector Augustus sent a complaint to King Frederick, who immediately summoned to his castle all Copenhagen pastors, all endowed professors of the University of Copenhagen (including Hemmingsen and Pratensis), and the bishop of Roskilde to answer the elector's charges. Peder Oxe was one of the three commissioners who examined them. Hemmingsen presented an eloquent defense of the peace and unity of belief and religious practice of the Danish church, describing German theologians as leaping about like cooks trying to please the palate of whatever noble they served. If Denmark paid attention to them, the result would be similar confusion. The difficulty was not immediately settled, but Peder Oxe told Hemmingsen privately not to worry, and matters quieted down.

By the end of December 1575, Tycho had returned to Denmark and was preparing to liquidate his assets for the move to Basel. The court had moved to Sorø Abbey for Christmas. King Frederick relished the rich food, mead, Rhenish wine, and ale of the abbot of Sorø, as well as the learned, amusing conversation around the abbot's table. Tycho went there to pay his respects to the king, report on his journey, and visit his aunt and foster mother Inger Oxe, who was at this time the noblewoman in charge of Queen Sophie's chamber.

Tycho had just turned twenty-nine and was an experienced courtier, polished by his travels and attendance at many courts. Garbed appropriately with flowing cape, feathered hat, and sword, he was an imposing figure, barrel-chested, elegant, and of distinctly noble bearing. His eyes were light-colored, and his hair, beard, and substantial mustache were reddish blond. In portraits, his false nose looks a fairly successful imitation, close to flesh-colored—though an astute portrait painter would have made it so in any case.

Tycho told the king about his visit to Wilhelm IV in Kassel and glowingly described this ruler who surrounded himself with scholars and artists. For work on Frederick's Elsinore building project, Tycho recommended the landgrave's former hydraulics expert, a portrait painter from Augsburg, and a sculptor trained in Italy, all of whom were willing to work in Denmark. He also reported on Rudolph of Hapsburg's coronation. He did not mention any intention of emigrating.

Tycho Brahe in a 1586 engraving by Jacques de Gheyn.

King Frederick regarded the impressive young astronomer with heightened interest. An emissary from the astronomy-loving land-grave had already informed him of Wilhelm's high regard for Tycho and recommendation that Frederick encourage him, and Frederick could read between the lines that if he did not, Wilhelm would. Perhaps there were rumors at court that Tycho had plans to move abroad. In any case, Frederick received Tycho with extraordinary gra-ciousness and offered him not just one but a choice among four fiefs. Two were castles on Baltic islands, rather far from Copenhagen but of great strategic importance. The others were Helsingborg, which Tycho's father had commanded, and Landskrona, both of which guarded the Øresund, the sound that led to the Baltic. At any of these castles Tycho would have power over hundreds of peasants and villagers, with servants, knights, troops, and courtiers to serve him.

But Tycho surprised the king by accepting none of these castles and major royal fiefs. Instead, he politely insisted that he needed time to think about them. Frederick had nothing better to offer. Nor was this a matter that he could simply let drop. He had no wish to lose Tycho to a foreign ruler or university. Moreover, it was a matter of honor that he show some form of substantial recognition for this young member of one of Denmark's most powerful families.

However, while the king was pondering the problem, Tycho wrote to Pratensis, "I did not want to take possession of any of the castles our good king so graciously offered me. I am displeased with society here, customary forms and the whole rubbish. . . . Among people of my own class . . . I waste much time."

In the two months following the king's offer, Tycho went ahead with preparations for his departure, but he was still somewhat ambivalent about it. To maintain an aristocratic lifestyle in Basel, he would have to purchase an estate, and the only way to afford that was to sell his portion of Knutstorp. That was a complicated undertaking, because he shared ownership with his brother Steen, his mother held a lifetime interest in the estate, and both lived there. Furthermore, refusing the king's magnanimous offer was sure to shed a bad light on the entire extended family. On the other hand, if his decision was going to be swayed by consideration for others, he should consider Kirsten first. There were heavy social duties connected with being lord and lady of a castle. Kirsten was as ill equipped as he was unwilling to perform these duties.

Tycho's uncle Steen, of Herrevad, not pleased at the prospect of seeing his favorite nephew disappear over the European horizon, took matters in hand himself. Steen found a roundabout way to let the king know that Tycho was considering emigration and what the reason was: The royal fiefs the king had offered involved duties that would distract Tycho from his work. As the king later told Tycho, it was at this point in the conversation that he recalled Steen mentioning, a year earlier, that there was an island in the Øresund that had a special appeal for Tycho. Frederick's next offer was designed to be one Tycho could not refuse.

The king had a flair for the dramatic. Tycho described his summons in an excited letter to Pratensis:

> Hear now what has happened these last few days . . . and hear it alone—do not reveal it to a soul, except our friend Dançey, when you two are alone. As I lay awake in bed, early on the morning of February 11, restlessly considering to myself the journey to Germany time and again from all sides and figuring out how I should be able to disappear without arousing the attention of my kinsfolk, when lo!—it was announced quite unexpectedly that a royal page had arrived here at Knutstorp, who had hastened the whole night through in order to bring me a letter from the king without delay (it was still dark of night, towards the break of dawn, and the sun did not rise for another two hours). Therefore I bade the page, a nobleman and a kinsman, to step up to the bed. He straightaway produced a letter and said that he had been commissioned by his king to ride night and day without rest, seek me out wheresoever I might be found . . . personally deliver the letter to me, and return immediately. . . . I broke open the letter and found that the king had commanded me to come to him without delay. This I did obediently without wasting a moment so that I presented myself before the king at the [royal hunting lodge of Ibstrup, in the forest about a mile from Copenhagen] that same day before sundown. Through his chamberlain, Niels Parsberg, he let me be called to him in private.

Frederick received Tycho with the news that his plans to leave Denmark were no longer a secret. The king said he also knew, and sympathized with, the reasons Tycho had not accepted his offers. He, too, was concerned that political and social duties should not interfere with Tycho's research. Frederick described a recent visit to Elsinore, where he had been overseeing the progress of the new castle: As he surveyed the seascape from one of the windows, his eyes happened to fall on the little island of Hven, on the distant horizon to the southeast—a beautiful, isolated place, not held by any noble in fief, carrying with it only minimal administrative obligations. Were Frederick to pay, out of the royal coffers, all expenses for building a suitable residence there and for founding and maintaining a research establishment, there was surely nothing abroad that could possibly lure Tycho away. His work at Hven would redound to his own credit and that of his king and his country. Tycho should not answer immediately, Frederick insisted, but consider the matter and reply as to whether he would accept this new offer.

Tycho returned to Knutstorp the next day, stunned by the proposal but still undecided. Another day passed before he wrote the letter to Pratensis asking for his and Dançey's advice. Both responded enthusiastically on the same day that Tycho's letter arrived. They emphasized, among other things, that the king's generosity was a powerful vindication for Tycho over those relatives and fellow nobles who disapproved of his straying from their own career paths and who had predicted a dismal future for him.

Frederick's new offer implied endorsement not only of Tycho's unorthodox career choice but also of his alliance with Kirsten. One of Tycho's later students reported that the king's "intervention" put to rest hard feelings toward Tycho among relatives who suffered diminished social esteem because of Kirsten's low birth. The gift of Hven was a clear indication of Frederick's favor and also (though the intention remained unspoken) a provision from the royal coffers of a haven for Kirsten and their young family.

By February 18, six days after Frederick made the offer, though Tycho had not yet decided whether to accept Hven as his fiefdom, he had made up his mind to stay in Denmark. He pledged his fealty to the Danish crown in the same words Brahes, Billes, and Oxes had uttered for generations as they began their service to the kings of Denmark. His royal pension began, nearly doubling his yearly income.

5

THE ISLE OF HVEN

1576–1577

FOUR DAYS LATER Tycho rode from Knutstorp to the harbor at Landskrona, on the eastern shore of the Øresund, and set sail across the icy sound to the island of Hven. The crossing took two hours. Cliffs on all sides of the island made the top in many places unreachable from the rock-strewn beaches. From Tycho's small boat these cliffs would have loomed tall as he drew near, and as he wrote in a letter to Pratensis, "Since the waves of the sea surround [Hven] on all sides, it has difficult, often quite dangerous landing places." Nevertheless, the boat managed to land on the north shore of the island, where there was a break in the cliffs and it was possible to get to the top.

It was a stiff climb, as well as cold, for this was late February, when the short winter days begin to lengthen at this latitude but bitterly sharp winds blow across the island. Tycho soon arrived at the only settlement, the village of Tuna. The cottagers there, ignorant of the future and the havoc he would wreak in their placid lives, very likely welcomed this aristocratic stranger with a warm fire. Tycho probably went beyond the village across the level top of the island to its center. From there, the promontory where King Frederick was constructing his new castle was visible on the distant horizon across the water. Tycho must have thought on that first day that this center point of Hven would be a splendid place to erect his own palace. He spent that night on the island and observed a conjunction of the Moon and Mars in the foot of Orion—his first recorded observation from Hven, February 22, 1576.

Tycho followed the king's suggestion that he take his time deciding whether to accept Hven or choose another fiefdom. He spent much of the next three months there, studying the conditions, weighing advantages and disadvantages. He measured the island by striding along the perimeter at the clifftops, counting 8,160 strides. For a man who had been considering emigrating to a warmer part of Europe, the climate was not enticing, but Tycho was a Dane by birth and breeding and accustomed to enduring cold and wet for much of the year. A more serious problem was that at Hven's latitude he would see less than he wished of those parts of the sky most interesting to an astronomer. The planet Mercury would frequently be out of sight below the southern horizon. Nevertheless, as spring damp and mists gave way to the promise of a hot, idyllic summer, Tycho was falling in love with Hven.

The island's isolation weighed both ways in his considerations. He would be near enough to Copenhagen to allow the occasional appearance at court or university, yet Hven was, as he put it, "free from the commotion of the common herd." Great sailing vessels, after queuing up to pay the toll across the sound at Elsinore, swept on in a steady procession as though the island were invisible. His time would be his own, his work interrupted only when he chose. But Tycho also realized that Hven's isolation made it less than ideal for a major building and landscaping project. He cast an appraising eye over the residents of Tuna, a seemingly docile lot. As their lord, he was entitled to two unpaid workdays per week, from sunup to sundown, from each farm on the island, and a certain amount of "cartage." That seemed very little in view of the undertaking he had in mind. Besides, they were not skilled laborers. Those and most supplies would need to come from the mainland.

By the time three months had passed, Tycho had made his choice. Set in a sparkling sea with a haze making the passing ships and the distant shores seem a mirage, his island's fields, pastures, village, and its tiny church of St. Ibb's glowed in the piercing sunlight of a northern early summer. Frederick had offered Tycho a paradise.

There were violent legends surrounding the early history of Hven, but more recent years had been tranquil. Until Tycho arrived in 1576, there had been little to break the peace and no lord to rule the island since 1288, when Viking marauders, led by the vividly named Eric the Priest-Hater, paused there to demolish several castles. The rulers who lived in them either perished or soon left. Traces of four fortresses were all that remained when Tycho came.

Several years later, after Tycho had erected a palace on Hven suited for entertaining royalty, Frederick's young Queen Sophie was obliged to spend an extra night when a violent storm made the Øresund impassable. The company gathered around the fire, and one of Tycho's students entertained them by recounting tales from before the time of Eric the Priest-Hater. Nordborg Castle, whose ruins guarded the landing where Tycho first came ashore, had once been a formidable stronghold. Lore had it that Lady Grimmel, whose family ruled the island, invited her two brothers to a feast there and murdered both of them during the festivities. Her maiden-in-waiting Hvenild was already pregnant by one of the brothers and gave birth to a son, Ranke, the true heir to the throne. While Ranke grew up, Lady Grimmel continued to rule from her four castles, Nordborg, Sönderborg, Karlshög, and Hammer. But Hvenild did not let her son forget his aunt's treachery and his father's fate, and when Ranke reached manhood he cast the evil Grimmel into a dungeon and abandoned her to starve. The name Hven was said to come from Hvenild, Ranke's mother.

The peasants also told of "Lady Grimmel's treasure," buried in the alder fen and guarded by a dragon (who, report had it, was only seldom seen). Supposedly, two golden keys in the sea could unlock the treasure, and at one time two boys saw them gleaming through the water. One of the boys revealed the secret, but when others ran to look, the keys had vanished.

More scholarly accounts report Stone Age habitation on Hven, a name evidently already used for the island in the ninth century, and it seems to have been a hive of activity during the Bronze Age. The sea battle of Svolder between powerful Viking forces took place at "Sandevolleön," the Viking name for Hven. The island's cliffs and location in the center of the sea-lanes made it a natural citadel, but after Eric the Priest-Hater's onslaught it was never again fortified.

With Tycho's coming in the spring of 1576, the little island was about to emerge from this murky history that was little more than legend and, in spite of its out-of-the-way location, move to the center stage of Europe. Yet Tycho would come to seem to the villagers of Hven, and to the descendants to whom they passed on the stories about him, not the enlightened genius of the age but a figure as mysterious and malign as the ancient Lady Grimmel herself.

On any of the other estates the king had offered him, Tycho would have found peasants, however disgruntled, accustomed to serving a lord of the manor. The peasants of Hven, nestled in their thatched, half-timbered village, had never experienced anything of the sort. They had been enjoying their utopian isolation without the interference of a lord for almost as long as their history could recall. For generations, forty peasant families had tilled the land that could be tilled, grazed a few animals on the broken areas where the cliffs had fallen away and other areas that did not lend themselves to tilling, fished, and managed to eke out a living, a portion of which they paid to a provincial governor on the mainland. They thought they were freeholders, that they owned their land. No one had disabused them of this notion.

The village's three great fields covered most of the northwestern

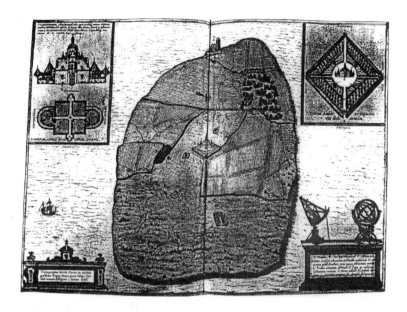

An old map of Hven that is not oriented strictly north to south. The village's three great fields covered most of the northwestern half of the island, nearest the village; St. Ibb's Church is visible at the top of the map.

half of the island, near the village, with common grazing land and meadows taking up the center and the southeastern half. There were almost no trees except for a grove of hazelnuts and the alders that grew in a moist area or fen, where the dragon was supposed to be guarding the treasure. St. Ibb's Church stood on the cliffs, and on another high place near the village there was a great windmill. All the islanders lived in the village, and with the exception of the pastor, a miller, and almost surely a blacksmith (though that may have been a part-time job combined with farm work), they were all peasants and rural laborers.

The islanders had been accustomed to governing themselves. Though their claim to be freeholders was supported by tradition, not by written documents, each farmer believed that he owned an individual narrow strip within the great fields. Since it was more convenient and productive to farm them together, they had organized a guild to set down bylaws and administer daily activities, choosing one prominent villager to be responsible for maintaining the peace and collecting annual taxes. The choice had to be approved by the provincial governor on the mainland, but Hven was remote and uninteresting enough to keep him from taking much notice beyond that.

In other ways Hven was more in contact with the rest of the world. Some islanders sold their produce in towns along the coasts of the Øresund. Others took their swine to graze in summer on acorns in the forests of Skåne, on the eastern side of the sound, where a few islanders had intermarried with local peasant families. But compared with communities on the mainland, Hven was remarkably independent in its day-to-day existence, which unavoidably set the stage for considerable disruption and even flat refusal to accept the situation should a lord of the manor suddenly present himself.

Tycho, after considering the advantages and disadvantages of various building sites, chose the exact center of the island, regarded by the villagers as common grazing land. It was his legal right to make this choice without consulting them.

On May 23, 1576, Frederick formally granted Tycho the island. Tycho sailed again to Hven. He and his party met the bailiff, "grands," and most of the population at an area near the center of the island, marked with boulders since ancient times as the meeting place of the "Hundred Thing," the local court of law. Tycho's clerk read aloud the parchment document that bore King Frederick's seal. The island was granted in lifetime fee to Tycho Brahe of Knutstorp, "to have, enjoy, use and hold his life long, and so long as he lives and desires to continue and pursue his *studia mathematica*." The grant required Tycho to "observe the law and rights due to the peasants living there, and do them no injustice against the law, nor burden them with any new dues or other uncustomary innovations." For the cottagers, everything would depend on who defined what was "customary," and there was little doubt who that would be.*

Soon after the formal granting of Hven to Tycho in May, the peasants began to find out how drastic a change was in store for them and their island. Instead of paying the accustomed taxes levied in the kingdom of Denmark, they were to render their labor services to Tycho, with no payment in return. Tycho's bailiff began taking down their names and noting the sizes of farms and cottage holdings so as to assess what Tycho's rightful dues of labor were from each family. The islanders soon learned about the obligation of each household to provide two man-days of labor each week and to appear with draft animals and wagons (the "cartage" requirement) on a prescribed number of days each year.

This was the peasants' role within the system, new to Hven but customary elsewhere. It was, in the theory of the time, not far different from Tycho's obligation on a higher rung of the ladder to serve the king. But the islanders were incapable of recognizing any parallel between their duty to sweat and toil for nothing and Tycho's duty, in rich robes and plush surroundings, to peer at the sky and boil up wizardly mixtures. As for Tycho, he surely no more thought of himself as placing unreasonable demands on his tenants than he thought of himself as placing unreasonable demands on his garden to produce plants.

Tycho spent much of the early summer of 1576 contemplating the site he had chosen for his palace, pacing it off, studying architectural theory, poring over designs, drawing circles and squares in harmonious proportion, setting stakes in the ground, pulling them out again, and repeatedly consulting experts and friends. During his re-

---

*This grant was to be only the beginning of Frederick's largesse to Tycho. He would later add the fief of Kullen, eleven farms in Skåne (where the Hven peasants grazed their swine), and the entire district of Nordfjord in Norway.

cent sojourn in Venice and its environs he had had opportunity to become well acquainted with the architecture of Andrea Palladio. Soon after their publication in 1570, Palladio's *I Quattri Libri Dell' Architettura* (Four Books of Architecture) had become a sensation throughout mainland Europe and England. They set the standard for years to come for the architecture of palaces and great houses. In addition to being a guide to classical architectural theory, the books could almost be used as a do-it-yourself manual, with clearly written text keyed to examples, detailed drawings, and ground plans. Palladio had combined a passion for the classical past with a gift for reinventing it. Tycho, in turn, reinvented Palladio.

At first glance, there was little resemblance between the palace Tycho designed and Palladio's simple, airy Italian masterpieces. It was the ideal of symmetry in Palladio, and the extension of this symmetry into the landscape, that most captured Tycho's imagination. In Tycho's house, in the projections and the towers that adorned it, and in his flower beds and orchards, he followed Palladio's example of using only pure geometric shapes.

In Palladian architecture, the symmetry of a house went beyond balancing architectural elements such as rooms, towers, windows, and avenues. More subtle proportions reflected a musical symmetry that had been an ideal since the Pythagoreans in the sixth century B.C. studied the sounds made by vibrating strings and discovered that there are harmonic ratios in nature. Johannes Kepler would later study those same relationships when trying to discover the design of the universe.

Tycho's house plan* had portal towers on the east and west sides of the house, each fifteen Danish feet wide and fifteen feet long.† The height of the facade was thirty feet (twice the width of the portal towers), the peak of the roof forty-five feet, the side of the central

---

*In his book *The Lord of Uraniborg,* Victor Thoren has provided these specifics about the way Tycho carried out this ideal of symmetry in the design for Uraniborg.
†The Danish foot that Tycho used was 259 millimeters, a little more than 10 inches.

An imaginative painting of Uraniborg, done by Henrik Hanson in 1862, is based on a sixteenth-century woodcut. It makes the house appear somewhat darker and more oppressive than it probably actually was. One of the observatory roof sections is shown open on the right.

square sixty feet. Fifteen, thirty, forty-five, and sixty form a ratio of 1:2:3:4. The progression 1:2:3:4 contains all the ratios of harp string lengths that since ancient times had been known to produce sounds pleasing to the human ear. Similar ratios of musical harmony were to underlie the proportions of Tycho's rooms and other relationships among elements of the building. Without knowing the designer's intentions, it would be all but impossible for someone viewing the house to recognize all these mathematical/musical subtleties, but Tycho was convinced that they would inevitably make his home and landscape a harmonious whole, pleasing to the eye, conducive to peaceful, intelligent pursuits, and inspiring to any sensitive person.

Tycho designed his castle-observatory to be a miniature gem of a palace, not nearly so massive as most noble dwellings. The entire building was no larger than even one of the four wings of Knutstorps Borg, and there were many larger castles being constructed by Tycho's contemporaries. Nevertheless, it was a major building project, and in late spring the grumbling heads of peasant households began sorting out which younger brother, younger son, or not-so-able hired hand could be spared with least inconvenience to fulfill the labor obligation. Poorer cottagers had no choice; they went themselves. They shouldered spades and trundled wheelbarrows and wagons to the center of Hven at sunrise, worked on the excavation, and kept at it until the sun set. The group changed constantly as men rotated in and out, fulfilling each household's obligation of two man-days of labor in a week.

There were no trained masons, carpenters, stone carvers, or tile workers, and certainly no hydraulics engineers among the local peasantry, so Tycho sought out and hired more skilled labor. For the most part he didn't have to look even as far as Copenhagen. King Frederick, on his own building site across the sound, was employing not only Danish craftsmen but Dutch and Flemish artisans as well. Tycho managed to convince the king or his architect that George Laubenwolf of Nuremberg was the only man alive with adequate skills to build the fountain that the king wanted for the central courtyard of his castle. While Laubenwolf was close by, just across the sound, Tycho also engaged him to design a water system for the palace on Hven. In an astounding innovation, water would not only supply a magnificent fountain in the lower central hall but would travel through "pipes reaching in all directions, to the various rooms, both in the upper and lower story."

Meanwhile, while Tycho was overseeing the work on his house almost on a daily basis, he still attended to the rest of his extensive new holdings. That first summer, he was constantly on the move between Skåne, Sjaelland, and Hven, taking boats across the sound and riding back and forth with a retinue across country, on horseback or by carriage, enjoying his new role.

As the plans for his palace took shape, Tycho's friends were almost as excited as he. Dançey said he would supply the cornerstone. Pratensis began to organize a ceremony for putting it in place and to draft an appropriate inscription. Dancey ended up writing the inscription himself, for tragically, in June, Pratensis collapsed and died while giving a lecture at the university. Tycho grieved for this friend who had encouraged him so enthusiastically when he had faced life with far less certainty than he did now, and who had shared many of his happiest moments.

Tycho studied planetary positions to find what date would be most propitious for laying the cornerstone, and when the ceremony had to be delayed, he studied them again. He chose August 8. High government officials, professors from the university, and Tycho's noble relatives arrived the day before. Tycho had commandeered hospitality for his guests in the homes of the beleaguered villagers of Tuna so that everyone could gather at the building site at sunrise. Late summer was perennially the most beautiful time of year on Hven, when at dawn mists hovered over the fields and the trees made long shadows that touched the horizon. In any direction Tycho and his guests looked, they could see the bountiful crops and grasslands of Tycho's domain, with the sea glowing beyond. It was not such a wild landscape as might be expected in the middle of a northern estuary like the Øresund, at least not at daybreak on a halcyon day in August. It was a place of soft, shimmering, tranquil beauty and of great expanse. The inscription on the stone that Dançey cemented into place and "consecrated with wines of various kinds" dedicated Tycho's future palace to the contemplation of philosophy, especially astronomy, and named it Uraniborg, the castle of Urania, muse of astronomy.

Only the absence of Pratensis darkened that golden day, but the tragedies of that summer and autumn did not end with his death. In September, Tycho's and Kirsten's older daughter, Kirsten, nearly three, died in an epidemic in Skåne. They buried her in the church at Helsingborg, and Tycho moved his wife, who was pregnant again, and the baby Magdalene to a more remote region farther north on the eastern shore of the Øresund.

In the late autumn, when building work slowed with colder weather, Tycho himself continued to live on Hven, but not in the new palace, which wasn't far enough along to be occupied. On December 14, his thirtieth birthday, he wrote in his journal that he was making his first observation of the Sun from the island. He also recorded observations on the twenty-fourth and twenty-fifth of December, so he spent Christmas there. Kirsten and Magdalene must not have been with him, for Kirsten was in Väsby, higher up the coast, on January 2, when their first son was born. Tycho never saw the baby boy, who lived only six days. The gravestone at Väsby called him the "natural son" of Tycho Brahe, as young Kirsten's bronze plaque, still in existence at Helsingborg, calls her his "natural daughter."

During that winter of 1577, Tycho began for the first time to make systematic observations from Hven. Ten to fifteen times a month, he would go out either at noon for an observation of the Sun or in the evening for positions of a planet. His astronomy was interrupted in late May by the christening of the long-awaited heir to the throne. Tycho retrieved his court attire from Knutstorp and with a full complement of servants—but without Kirsten—went to join in the more than two weeks of festivities. His mother, his uncle Steen, and his brother Steen were godparents, but the king and queen gave Tycho an enormous honor and responsibility as well. They asked him to cast the horoscope of the infant Prince Christian, to discover what the stars promised for him and the kingdom.

It was not an assignment to be taken lightly. If it was accurate, the horoscope would be an extremely valuable document, making it possible for the prince to anticipate personal and political crises. Tycho began by using both the Prutenic and Alfonsine Tables to calculate the positions of the planets at the moment of the prince's birth. Then he employed the observations he had made during the past winter to correct the positions for the Sun, Venus, Mars, and Jupiter, only relying directly on the older tables for Mercury, Saturn, and the Moon. With these and further calculations, Tycho already had twenty-seven pages. The royal family was to have a state-of-the-art horoscope.

Tycho predicted, with plenty of supporting astrological reasoning, that Christian would be "well-formed, righteous, charitable, nimble and capable of hunting and warfare, and equally nimble of mind for a broad spectrum of cultural and intellectual interests." The less welcome news was that the young man would be a little "over fond of sensual pleasure, subject to danger when it came to religious matters, would have to overcome adversity to win honor and riches," and would have few children. This second part of the horoscope added another forty-four pages.

Tycho pointed out that even if an astrologer produced a superbly accurate horoscope, as he was certain he had done, an error of as little as four minutes in the royal clock establishing the time of birth would render the document useless. He also repeated what he and Hemmingsen had agreed on at the time of the lecture in 1574, that it was possible for either divine intervention or human free will to avert the fate predicted by the stars. (Presumably that was how Christian later managed to have eighteen illegitimate children.)

Tycho had finished the horoscope by the end of June and translated it into High German, the language with which Queen Sophie was still most comfortable. He dated it July 1, 1577, and soon thereafter presented it to the king at Kronborg Castle at Elsinore.

The second year of construction on Hven was well under way, with the villagers of Tuna again laboring every day but Sunday. With the foundations completed and more skilled workers taking over as the walls rose, Tycho set the peasant laborers to digging no fewer than sixty fishponds, filling them with water, trundling the excavated earth and rock over to build the perimeter wall for his garden, leveling the area within the wall, and planting the first trees and shrubs. Tycho's tenants were particularly displeased with the job of carting and planting all the trees, which struck them as a frivolous and tedious bit of make-work. Others of the men were sent to Kullagaard on the mainland to cut wood and boat it back to Hven to fuel fires for brick making, for Tycho had noticed the potential of the great forests in Kullen when he had taken Kirsten and Magdalene north to escape the plague.

With all this work going on and the islanders increasingly disgruntled as no end seemed in sight, the bailiff, overseer, and "summoner" on Hven, who knew the villagers well, began to notice that faces were missing. Younger men with no wives or children to tie them to Hven were finding ways to escape the island. The desertion of some of his most able workers troubled Tycho, and he was not about to allow it to continue, but at this stage of his life he was finding it possible to take such management problems and exasperation in his stride and still not neglect his astronomy. He had a new and much-improved quadrant, and with it, among other observations, he recorded lunar eclipses in April and again in September.

Later in that autumn of 1577, the skies provided a far more unusual and mysterious spectacle, one of the most thrilling events of Tycho's career as an astronomer. In the early evening of Wednesday, November 13, Tycho was out in the gathering dusk catching fish for dinner in one of his new ponds. Looking toward the west across the fields of his island, he saw an exceptionally bright star. The only planet in the evening sky at the time was Saturn, and Saturn was never so bright. Fish and dinner were forgotten, and Tycho watched, transfixed. As the sky continued to darken, the star grew a long, fiery tail. A comet! Ever since the nova five years earlier, Tycho had been longing to see a comet.

6

WORLDS APART

1571–1584

THE COMET REFLECTED in Uraniborg's ponds in the autumn and early winter of 1577 hung in the skies of all Europe and far beyond. Thousands of awestruck men, women, and children went out at night to peer at it with curiosity and superstitious fear, wary of what this apparition would do and what it meant. One clear evening in southern Germany, a small boy named Johannes Kepler clasped his mother's hand and followed her up the hill above the little town of Leonberg, where she pointed out the bright star with a tail. He did not see it clearly, for his eyesight was poor, and he was too sleepy for the comet to make much of an impression. More unusual and remarkable, for him, were the warmth and companionship of the moment with his mother—an exceptional instance of grace in a harsh, dreary childhood.

Johannes Kepler was five years old when he saw the comet. He had been born on December 27, 1571, in his grandfather's house in Weil der Stadt, a small city on the edge of the Black Forest near Stuttgart. The time of his birth was two-thirty in the afternoon, a detail that was carefully written down even in this disorganized household, for in this era astrology was still a respected discipline. Tycho

Brahe had celebrated his twenty-fifth birthday earlier the same December.

The Keplers had once been a noble family. At Whitsuntide in 1433, Emperor Sigismund had bestowed a knighthood on Johannes's great-great-great-great-grandfather for valiant military service at the Tiber Bridge in Rome. By the time his descendants had emigrated to Weil from Nuremberg, about fifty years before Johannes's birth, straitened financial circumstances had brought them down to the level of craftsmen, still cherishing tales of better days and a family coat of arms.

Later, when Johannes was in his mid-twenties, he drew up a "birth-horoscope" of his ancestors, and the notes he made for that describe his grandparents and parents and some of the events of his childhood, including seeing the comet. Kepler was usually respectful and loyal in his treatment of his relatives. However, in these fragmentary jottings that he made solely for his personal use and never intended for publication, he was devastatingly candid about his severely dysfunctional family, as well as about himself.

The Keplers evidently maintained a reasonably good public image. Grandfather Sebald, head of the family, had been bürgermeister of Weil der Stadt for ten years when Johannes was born, and a portrait of him shows a well-dressed, distinguished, bearded man with a ruddy complexion. Johannes's notes said that though Sebald was not eloquent, he gave wise counsel in the city and had a strong enough personality to see that his opinions were respected and his advice heeded. This proud civic figure did not, however, come across so well in the privacy of his home. Kepler described him as arrogant, stubborn, sensual, and irascible, with little affection for the grandchildren who spent their early years underfoot in his house. "His face betrays his licentious past," wrote Kepler.

Johannes gave an equally unsympathetic picture of his grandmother. She was a restless woman, thin, fiery-tempered, resentful, clever, "blazing with hatred," "violent, and a bearer of grudges," and a liar. She was also devoutly religious.

Johannes's father, Heinrich, was their fourth son, and he, by his own son's report, was a vicious, immoral, brutish, uneducated man. "He destroyed everything. He was a wrongdoer, abrupt, and quarrelsome," and he "beat his wife often." Theirs was "a marriage fraught with strife." Through a combination of bad behavior and bad luck, Heinrich had brought the Kepler family to an unprecedented low. Before Johannes was three, Heinrich set off adventuring and fighting as a mercenary. He returned only occasionally to his wife and children, and his short stays were not happy.

The task of raising Johannes and his brothers and sisters—there were seven children, four of whom survived to adulthood—fell mainly to their mother, Katharina. She was the daughter of another prominent civic leader, the bürgermeister of nearby Eltingen, Melchior Guldenmann, who was also an innkeeper. Kepler described his mother as small, thin, dark-complexioned, garrulous, quarrelsome, not a pleasant woman. Her acquaintances regarded her as an evil-tongued shrew.

Young Johannes, the eldest of the children, resembled Katharina in appearance. They were also alike in having restive, inquiring minds, but Katharina had no education, and her interests were herbs and homemade mixtures for healing. What in her son would develop into a rich intellectual curiosity was, in her, often only nosiness. When Heinrich was at home, she responded with pouting and stubbornness to his harsh, rude treatment. "She could not," wrote Kepler with pity, "overcome the inhumanity of her husband."

Kepler also described aunts and uncles and some cousins who lived in the house in Weil der Stadt. Among them were Uncle Sedaldus, who was "an astrologer, a Jesuit, acquired a wife, caught the French sickness, was vicious," and Aunt Kunigund, who was poisoned and died.

In the spring of 1575 Katharina Kepler left three-year-old Johannes and his infant brother in the care of these relatives and went off to follow her soldier husband Heinrich. In her absence Johannes nearly died

A copperplate drawing of Weil der Stadt, the town where Kepler was born.

of smallpox, probably the illness that impaired his vision. The prodigal parents returned after a year.

With both nature and nurture decidedly against the two Kepler brothers, their future looked bleak. Johannes was puny and weak-sighted. Heinrich, two years younger, was an epileptic. Johannes recalled that Heinrich was beaten roughly, and animals frequently bit him. He nearly drowned, nearly froze to death, nearly died of illness, and ran away from his apprenticeship to a baker when his father threatened to "sell him." After that he appeared only occasionally at home, much as his father did, often returning bruised and broken, robbed of everything he had, making his way back by begging. All his life—he died at the age of forty-two—his mother considered him the bane of her existence.

The younger Kepler children turned out better. Margarethe, a gentle, sympathetic girl, later married a clergyman, and she remained close to Johannes and loyal to her mother even through the worst of times. The youngest surviving sibling, Christoph, grew up to be an honorable, correct man, though not so unfailingly loyal as his sister. He became a respected craftsman, a pewterer.

The little city of Weil der Stadt, surrounded on all sides by the duchy of Württemberg, nevertheless enjoyed the status of an imperial free city within the Holy Roman Empire and sent its own representative to the Imperial Diet. Though its name implied that the Holy Roman Empire was in some way the legatee of the Roman Empire, the standard and fairly accurate quip is that it was not holy, not Roman, and not an empire. Ruled in theory by the Holy Roman Emperor in Prague and the Imperial Diet, it was made up of many units—duchies like Württemberg, cities like Weil der Stadt, bishoprics, and other principalities—that today have become Germany, Austria, and the Czech Republic, as well as parts of Poland, France, and Holland and sundry other bits and pieces of Europe.

Under the 1555 Peace of Augsburg each local leader decided whether Catholicism or Lutheranism would be practiced in his domain. An exception was made for free imperial cities like Weil der Stadt: If both religions had previously been practiced there, both were allowed to continue. The duchy of Württemberg, which surrounded Weil, was by decree of its powerful duke officially and vehemently Lutheran. Weil itself was mostly Catholic, but there were Lutherans there as well. The Keplers were part of this Lutheran minority—a not entirely comfortable situation, though grandfather Sebald seemed not to have found it an impediment to political advancement.

In 1576 Johannes's father renounced the right of citizenship in Weil and moved his young family to Leonberg, not far away but part of Lutheran Württemberg. It was from the hill above that town that Katharina and Johannes viewed the comet.

❦

T Y C H O   B R A H E and other scholars, though not immune to the disquiet about the comet felt by less educated people like the Keplers, undertook to study it zealously from a scientific point of view. Tycho's first step was to write a careful description and make a drawing. The comet's head was seven to eight arcminutes in diameter and bluish white, the color of Saturn. Its tail was reddish, like a flame seen through smoke. To pinpoint its position, he measured its angular distance from two prominent stars. Tycho wanted to find out how far away from Earth the comet was, and that meant, as it had for the nova, finding out whether it displayed any parallax shift.

Looking for the comet's parallax presented new challenges. It was positioned too near the Sun to be visible except for an hour or so just after sunset, and that was too short an interval for Tycho's position on Earth to change sufficiently to provide any perspective on it. Furthermore, it was known that comets were in motion. With the nova, Tycho had been able to assume that *any* movement it displayed against the background was a manifestation of parallax. He could assume no such thing for a comet. Its change of position against the background would be partly, maybe totally, attributable to its own motion. Without knowledge of what that motion was, any attempt to measure parallax would be futile.

Partly due to cloudy weather and the wait for longer evening visibility as the autumn days grew shorter, it took Tycho ten days after he first saw the comet to determine that it moved an average of about three degrees of arc per twenty-four hours. Because every twenty-four hours brought Tycho back to the same viewing position, that motion could not be attributable to parallax: It had to be the comet's own motion. Three degrees of arc per twenty-four hours is about seven and a half minutes of arc per hour. Tycho next made observations three hours and five minutes apart, during which interval the comet's own motion—the seven and a half minutes of arc per hour—should have moved it, he calculated, about twenty-three minutes of arc. If it appeared to move *more or less* than that, the difference would be attributable to a parallax shift. Tycho found that the comet appeared to move only twelve minutes of arc. Parallax shift, he concluded, had to account for the other eleven minutes of arc

That result was disappointing. It was a borderline case whether this amount of parallax indicated that the comet was above or below the Moon. Further study soon yielded something more decisive. Tycho reconsidered the comet's daily intrinsic motion and found it closer to two degrees than three. He did the calculations again, and this time he discovered almost *no* motion left over to be accounted for by parallax. Additional observations near the end of December bolstered the case for the comet having virtually no parallax at all. Tycho saw the fading comet for the last time on the twenty-sixth of January, when the Moon, which had drowned out the comet's light for two weeks, had waned enough to allow him one last glimpse. He had already begun to write his conclusions.

Tycho felt he had settled the question of whether comets are closer than the Moon or farther away. Aristotle had been wrong about the "unchanging" heavens. This comet was a change, and it was indisputably beyond the Moon, though Tycho could not specify precisely how far beyond.

Tycho reached a second conclusion. The comet moved in the same direction the planets move, and this movement had for the first week carried it out very rapidly in front of the Sun, but after that it had moved more slowly and become dimmer, suggesting it was moving farther away from Earth. Then the Sun had begun to catch up again. Tycho concluded that the comet was orbiting the Sun.

He began to plan a book. He knew from previous experience that it was difficult to change the views of his colleagues and the public that followed them blindly. His new book needed to be more rigorous, more detailed, longer, than its competitors. While the comet was still visible, Tycho started a notebook of star observations so that he could use his own coordinates for reference stars to locate the comet rather than rely on the old catalogs.

The first order of business, however, was to file a private report to the king. It was only a little embarrassing that Frederick had seen the comet two days earlier than Tycho. Someone at Sorø Abbey, where the court was lodged at the time, had pointed it out. Tycho had also been preempted in his royal reporting by one Jørgen Dybvad, an open-minded man when it came to Copernican astronomy but preoccupied in the present case with what the comet portended. His pamphlet predicted bad weather, crop failure, religious troubles, pestilence, war, even that "the day of the Lord . . . is at hand." Dybvad was an ambitious and powerful figure at court and in the university and a potential competitor. Tycho saw an opportunity to best him.

His report to Frederick began with a description of the comet, including technical details that he promised to expand on in a later, more formal publication. This material could not have been of enormous interest to Frederick, but Tycho saw it as a way of reinforcing his image as an expert and giving greater credence to his interpretations as an astrologer. The next part of the report obviously referred to poor Dybvad: "Pseudoprophets who have thought [that comets might presage the apocalypse] and have mounted too high in their arrogance and not walked in divine wisdom will be punished." Tycho was not, however, ready to say there was nothing to worry about. Historically, he reminded the king, comets had always meant "great scarcity . . . many fiery illnesses and pestilence and also poisonings of the air by which many people lose their lives quickly . . . great disunity among reigning potentates, violent warfare and bloodshed and sometimes the demise of certain mighty chieftains and secular rulers." Because of its position in the sky and other characteristics, this comet was worse than usual and augured "an exceptionally great mortality among mankind."

With his reader now brought to the point of despair, Tycho advanced the more soothing and optimistic theme of the arguments he had made in his first university lecture—that heavenly events do not determine the future. Resorting to anguished prayer was not the proper course, he advised, for rational exercise of free will and appropriate action could change the effects of the comet. The king might even prepare in advance to reap the benefits if the prediction about the "demise of certain mighty chieftains and secular rulers" should turn out to apply to Ivan the Terrible of Russia. And Frederick would be wise to get ready for "Spanish treachery" if the comet had special "significance over the Spanish lands and their reigning lords." Tycho continued in this vein, knowing that these interpretations in terms of political policy would appeal to Frederick. When it came to demonstrating the value of astronomy and astrology to the king—and implying that it was an extraordinary advantage to have Tycho himself at the king's right hand rather than a defeatist like Dybvad—Tycho was playing it to the hilt.

The report also revealed a trend in Tycho's thinking—or perhaps it was merely rhetoric—that the king may not completely have shared. Tycho painted himself as a man who not only abhorred the politics of court life but also longed for peace and justice on a wider than personal scale. The comet, he wrote, might mean "well-deserved punishment for inhumane tyranny," and for "those who were associated with [violence and warfare], those who are always on the prowl [causing] great injury to others."

If the peasants on Hven had been able to read their master's hyperbole about "well-deserved punishment for inhumane tyranny," they might well have responded with rude noises, for during the same visit Tycho asked Frederick for assistance in dealing with the peasants who were fleeing Hven. These deserters, Tycho pointed out correctly but also self-righteously, were violating the law of villeinage and thus placing a greater burden on those who remained. The king's reply made it clear that the law that applied elsewhere also applied on Hven: Tenants could leave an estate only with the permission of their lord.

It was possibly also on this occasion that Frederick reiterated his promise to Tycho of the canonry at Roskilde Cathedral on the death of an incumbent. He could look forward to the incomes from endowments of the Chapel of the Magi there.

Tycho's ties to the king grew even stronger in the summer of 1578 as the construction progressed at both Uraniborg and Kronborg, the new palace at Elsinore. Craftsmen, materials, and architectural ideas moved swiftly back and forth across the sound.

For his "architect" Tycho had made an unlikely but inspired choice. Hans van Steenwinkel was a Dutch master mason who came to work for King Frederick at Kronborg. Tycho brought him across to Hven, hoping he might be trainable as a master builder. He gave Steenwinkel some instruction in astronomy and geometry, explained the symmetrical scheme of the building and grounds, and set him to work drawing more detailed plans. Steenwinkel was a quick study. Before long he had mastered classical and Italian Renaissance architectural theory as well as perspective drawing and was producing designs for windows, spires, domes, and other architectural details that pleased even the exacting Tycho. So great was Tycho's confidence in Steenwinkel that he put him fully in charge of the construction.

A few months after Steenwinkel came, Tycho engaged a twenty-three-year-old university graduate named Peter Jacobsen Flemløse to assist him in astronomy, alchemy, and other work—the first of a long procession of assistants and students that would finally end with Johannes Kepler. Flemløse, like Steenwinkel, was able and quick-witted. Tycho taught him to use the cross staff and the sextant and delegated to him the task of compiling a new catalog of reference stars for the comet. Flemløse also liked to draw, and from this time forward whimsical pictures adorned Tycho's star catalogs and observational journals.

When the building season slowed down once again in the autumn of 1578, Tycho's interest turned back to the comet and his book. He gathered all the observations he had made of distances from the comet to twelve stars, as well as descriptions of the observing conditions (the weather and other things such as moonlight that affected the observations), and put this material in the first chapter—an unusual way to begin in Tycho's day. It was unprecedented for anyone to share so much data with his readers. The book was also unusual in its author's willingness to admit error and his capacity to analyze why the error had occurred. On the first night of observation, the position of the comet figured from the twelve stars was at odds with the position figured from the Moon. Tycho left this discrepancy in the book, permitting his readers to see the conflict. Later, when printing was almost completed, he found the reason for the problem and added an "annotation by the author derived from later observations of the Moon."

With so much introductory material, it took Tycho ninety pages to get to the real crux of the matter: whether the comet was higher than the Moon. By now he was more convinced than ever that it was, and that the comet moved in a great circle, like the Sun, the Moon, and the planets, though with a less regular motion. He estimated that the comet came no nearer to Earth than six times the minimum distance of the Moon.

In December another royal prince was born, and that meant another horoscope, but Tycho was not averse to setting his book aside. There was already a plethora of commentaries on the comet coming off the printing presses of Europe. Tycho knew that he would have to produce something extraordinary to make an impact, and, curious about what his competitors were saying, he began collecting, through friends abroad, all the publications on the comet that they could put their hands on.

❧

AFTER HEINRICH KEPLER moved his family, including four-year-old Johannes, away from the overcrowded house in Weil der Stadt to Leonberg in 1576, he stayed with them only about a year before leaving again to sell his services to the Belgian military. Home became a more peaceful place for Katharina and her children. But Heinrich's Belgian adventure was a disaster. He lost what little fortune he had and nearly ended on the gallows. He trudged back to Leonberg and announced that they had to sell the house. The family moved to a rented property in Ellmendingen. After three years of near destitution, they somehow managed to acquire some property back in Leonberg and return there. It was at about this time that Johannes, reading of Jacob and Rebecca in the Bible, decided that if he should ever marry he would take them as a model. Their faithfulness was a marked contrast to his unstable and undependable parents. Five years after the move back to Leonberg, when Johannes was sixteen, Heinrich abandoned his family forever. Johannes never saw his father again.

Johannes would grow up an ardently religious man whose life was repeatedly, tragically disrupted by the political/religious strife around him. Nevertheless, at the start, the religious establishment served him well. The Lutheran Church's commitment to education provided a singular stroke of good fortune in an otherwise hopelessly bleak childhood. The Lutheran duchy of Württemberg had established a fine free school system, and this system rescued Johannes.

Much more information survives about Kepler's school days than about Tycho Brahe's. Johannes began at the German Schreibschule in Leonberg, where pupils learned to read and write the German they needed for everyday life. His teachers recognized an exceptional young mind and transferred him to a "Latin school." The dukes of Württemberg had established such schools in all small towns like Leonberg.

Johannes's transfer was a significant advancement, for Latin schools were the Lutheran substitute for the monastery schools that, before the Reformation, had provided primary education for boys who would become civil administrators, clergymen, and scholars. Latin was the common language in which educated men all over Europe communicated, lectured, debated, and wrote books; and Leonberg's Latin school set its boys firmly on this path by requiring them from the start to converse with one another day and night in Latin, or not at all. In the first year, they learned to read and write the language; in the second they endured endless grammar drills; in the third they read the classical texts.

It took Johannes five years to complete the three-year course. The move to Ellmendingen interrupted his education when he was about eight and had been a pupil in Latin school for a year. During that period of abject poverty, his parents set him to heavy agricultural labor rather than allow him to continue in school. Those two years were hellish for Johannes, for not only was he an undersized weakling of a child, pathetically unsuited for such work, but he loved school. His one source of happiness had been snatched away. However, when the family fortunes improved, his parents reenrolled him at age ten. Two years later, in 1584, he passed the competitive examination marking the end of Latin school and moved on to the "lower seminary" at Adelberg, where his room, board, and tuition again were free, courtesy of the duchy. After two years there he advanced to the "higher seminary" at the former Cistercian monastery at Maulbronn for two more years of study. Maulbronn was a preparatory school for the University of Tübingen.

It was at about the time that Kepler left Latin school to enter Adelberg that he first became aware of a potentially explosive rift in the Protestant world. He heard a sermon in Leonberg given by a young deacon who spoke vehemently and at length against the Calvinists. Twelve-year-old Kepler went away deeply worried about this harsh controversy between those who adhered to slightly different confessions of the same faith. At the age of thirteen, he wrote a letter requesting that the University of Tübingen send him copies of Martin Luther's disputations. Kepler decided to make it a practice, whenever he heard a preacher or lecturer argue about the meaning of the Scriptures, to consult the passages himself rather than to take anyone's word for what they meant. He usually decided that both interpretations had good points. This was not a healthy attitude for a young man who was hoping someday to find himself in a Lutheran pulpit—indeed for anyone wanting to survive unscathed in the political/religious milieu of the late sixteenth and early seventeenth centuries. Rather than win friends in both camps, it was likely to make enemies all round. As Kepler recalled his youthful inclination to see all sides of an argument: "There was nothing I could state that I could not also contradict." It was a gift, and sometimes a curse, that would remain with him all his life.

7

A PALACE OBSERVATORY

1578–1585

IN THE LATE 1570s and early 1580s, Tycho Brahe's fortunes continued to soar. The long-promised, prestigious canonry at Roskilde Cathedral and the incomes from the endowment from the Chapel of the Magi there were finally Tycho's when the incumbent canon died. The Chapel of the Magi was no ordinary chapel. It was and still is one of the most lavishly decorated in Denmark, housing a tomb, then under construction, for King Frederick's father, Christian III.

Though the earlier agreement had been that when Tycho received the canonry he would relinquish the fief of Nordfjord in Norway, he successfully appealed to the king to allow him to keep both, giving him a higher total income than any other scholar in Europe. King Frederick was setting a new standard of royal support for scientific research. He was also spoiling his young favorite, who was learning to think of himself as the equal of kings and to assume that his priorities would always be clearly recognized as the priorities of the kingdom.

Tycho's grandiose dreams for the island of Hven were becoming a reality. On the plot so carefully and symmetrically laid out at the center of the island, a magical structure was rising. Even though Tycho and his family would not occupy the house for another eighteen months and he wouldn't declare the building complete until the autumn of 1581, in July 1579 Tycho sent messages to his friends Vedel and Dançey that it would be well worth their while to pay a visit. There was ornamentation still to be done on the exterior of the house, and inside only the framing was finished, but the grounds were laid out according to plan, and it was possible to see the shape of things to come. Kirsten lived in temporary quarters on Hven that summer, and it was there that she gave birth to another daughter, Elisabeth.

"House" hardly sufficed to describe the magnificent vision that confronted Vedel and Dançey on the old common lands of Hven. In its geometric extravagance, Tycho's design rivaled Ptolemy's scheme of the cosmos. An earthwork wall surfaced with stone (built up of earth the peasants had dug to make the fish ponds) surrounded an 839-foot-square area. The plot enclosed by the wall formed a compass (see color plate section), with four avenues leading from the compass points to the center where the mansion stood. One of these avenues ran from the house to a two-story gatehouse at the east corner. An identical gatehouse mirrored it at the west corner. Later, kennels perched above these gates would house English mastiffs to announce the approach of visitors and frighten off intruders. The servants' quarters were at the north corner of the compound, and Tycho's printing establishment would later be situated at the south corner, with both buildings designed as miniatures of the main house.

Inside the earthwork wall were fruit-bearing and ornamental trees, eventually three hundred of them, each one a different variety. This orchard/arboretum enclosed another inner square, set off by a low wooden fence, with geometrically laid out beds for a botanical garden of flowers and herbs—as many interesting and exotic examples as could be brought to Hven. Wooden fencing also lined the four avenues and a central circle where the house stood facing east.

It was a palace like no other in the world, a whimsical bauble with carved sandstone ornamentation and a remarkable roofline noticeably lacking castle turrets or crenellation (see page 85). Crowning the building instead was a large pavilion with clock faces on its east and west fronts. At the roof peak, sixty-two feet off the ground, a smaller cupola housed the clock chimes. Two cone-shaped wooden roofs flanked this central block. They were the roofs of Tycho's primary observatories and were connected by galleries to smaller, similarly roofed projections in a hen-and-chicks arrangement. Pyramids, spires, dome, cupola, chimneys, galleries, and other fantastical decorations suggested more an illustration from a northern fairy tale than a Palladian villa. The appearance of Tycho's castle cannot have failed to reinforce the image the peasants had of him as a golden-nosed wizard brooding over their island, robbing them of their freedom, and performing strange incantations, experiments, and transmutations in his subterranean alchemical laboratory.

However, in addition to the underlying harmonic rationality of the design, there was other method in this architectural madness, and practicality rather than whimsy necessitated much of what Tycho and Steenwinkel built. The galleries made the primary and secondary observatories accessible to one another making it unnecessary to descend to the floor below. The chimneys ventilated no fewer than fourteen fireplaces and sixteen alchemical furnaces. The weather vane that topped it all off—a golden Pegasus—was connected with a mechanism so that the wind direction could be read from inside the house. The triangular panels of the cone-shaped observatory roofs could be removed individually to allow Tycho and his assistants to study one part of the sky or another.

Finally, in the early winter of 1580–81, Kirsten had a house to which she could "keep the keys." More than the immediate family moved in, for the building combined observatory, alchemical laboratory, and spacious family living areas, and also provided quarters for students and assistants. What Tycho had created was not only a self-sufficient manor of the sort that nobles had been building and fortifying since Roman times, not only a mansion that incorporated the architectural and aesthetic principles of the high Renaissance, but also something that could operate like a university professor's boarding house.

Tycho had woodcuts made of Uraniborg, including a floor plan, and wrote a description of what the house was like inside, all of which he later included in his sumptuous book *Astronomiae Instauratae Mechanica*. His floor plan showed a large room on the ground floor, which he marked *D*, that he indicated became the center of both family and scholarly life, where most of the dining and talking and even some of the research took place. In the summer, the focus of activity moved to a room on the floor above, with west-facing windows. In the first years, a corridor from the entry on the ground floor led through to the middle of the house. There, at the geometric center of Uraniborg, stood the remarkable fountain designed and engineered by Laubenwolf, with a rotating, hydraulically run figure that sprayed water into the air in all directions as it turned. Not only was it beautiful, but it also called attention to a feature of the house—running water—that even Tycho's contemporaries Queen Elizabeth I of England and Henry III of France could not boast of having at Hampton Court and the Louvre.

Sometime later, the wall between the entrance corridor and the room Tycho indicated as *D* was knocked out, so that the entry steps led directly into "the winter dining room or the heating installation." Since fourteen rooms had hearths and chimneys, Tycho probably meant that there was always a warm fire burning here in a huge tile stove, keeping it fairly comfortable for him, his students, guests, and assistants to read and study, even in a bitter Danish winter. The room* was probably wood-paneled with a beamed ceiling, its walls hung with paintings and shelves for books. Kirsten and their chil-

---

*The description of the room and of a meal that would have been served in such a setting comes from John Robert Christianson, in his book *On Tycho's Island*. Christianson, in turn, based his description on Tycho's own account and on accounts of similar rooms and dining practices in other Danish manor houses.

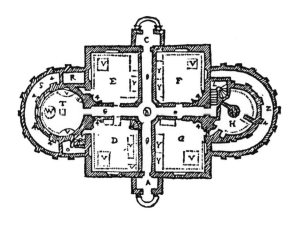

Figure 7.1: Tycho's floor-plan drawing for Uraniborg, from *Astronomiae Instauratae Mechanica.*

dren joined Tycho and his scholarly retinue here to eat the long meals that fused the ceremonial dining habits of the Danish nobility and a mealtime atmosphere more characteristic of a professor's household.

According to custom, a great oak table would have been placed so that Tycho and Kirsten (seated on the "high seat" in the middle of one side of this "high table") were flanked by their children, Tycho's assistants, and their guests—all arranged in order of rank. Those of less importance sat on benches along the walls on either side, with tables set before them too, leaving the fourth side of the room open for serving and for entertainers, who included a jester but often were the students, assistants, and guests themselves.

Though no record of a typical menu survives from Uraniborg, there is no reason to think that Tycho ate any less well than his Brahe relatives. The menu in another Brahe castle for a single meal included a first course of soup, carp, pike with turnips, venison with currant sauce, chicken pâté, goose liver with cucumbers, sugar cakes, and beef with horseradish; the second course was crabs and roast lamb with beets, followed by an almond sweet and a tart. It took time to eat a meal like this, with each course possibly lasting an hour or more, but it meant that once a day, for all his years at Uraniborg, Tycho had his staff and visiting scholars together for lengthy discussion of the work they were doing, as well as for enjoyment of one another's ideas, songs, stories, and theatrical skits.

Tycho and Kirsten probably often slept in this same room in winter. It was not the practice among those who lived in such castles in Denmark to have specified "bedrooms." Beds were portable. When the family were ready to retire, they told the servants where the beds should be set up, and it was done.

Across the corridor, in a room Tycho marked E on his floor plan, there was a large section on the south end of the west wall where Tycho would soon erect his magnificent "mural quadrant," and Steenwinkel would draw the architectural portions of the mural. F and G were guest chambers with, in Tycho's words, "desks for the collaborators." T, a circular library, would house Tycho's three thousand books and his great globe. The globe was important enough to rate its own designation as W on the plan. Above the library and reached by a staircase off the south wall of the room was the south observatory with its removable roof panels. Below the library by the same staircase was the subterranean alchemical laboratory, with sixteen furnaces of nine different kinds. Four of them were visible in niches around the walls in Tycho's elevation drawing (see color plate section). Later, Tycho moved some of his alchemical work even more into the heart of the house. He installed five furnaces in the dining room itself so that it was possible to keep an eye on lengthy distillation processes without having repeatedly to interrupt a meal to traipse down the basement stairs. The smells of alchemy mingled with the smells of the food.

The basement level extended a few feet beyond the rest of the house in all directions, and its windows were aboveground. In addition to the laboratory, it housed pantries, a wine cellar, a salt cellar, a wood cellar, and the deep well that supplied water for the water system. At the opposite end of the basement from the laboratory, another staircase led up to the ground-floor kitchens that balanced the library in the symmetrical plan of the house. So integrated into that plan were the kitchens (balancing the library, no less) that this may have been Kirsten's domain rather than an area for servants only.

On the floor above the ground floor, the central block consisted of more living accommodations. Looking toward the east were the "red" and "blue" chambers and a "yellow octagonal chamber" between them. Here was the Queen's Chamber, which Queen Sophie occupied when she visited, and the King's Chamber—though the king never stayed the night. The west half of the second floor was taken up by the fifteen-yard-long "summer dining room"; Tycho called it the "green room" because the ceiling was painted with pictures of plants. The windows looked over the great west wall to the Øresund and gave a view in the distance of ships passing through the tollgate at Elsinore. The areas above the library and the kitchens were the observatories whose roof panels could be opened and in which Tycho's instruments stood like fabulous sculptural masterpieces. There were galleries connecting the main observatories with the satellite observatories supported by single pillars (see page 85).

What the interior of the central block was like above this floor is less clear. Another flight of stairs spiraled upward. If it was a large open spiral leading up into the dome and the pavilion at the top of the house, the dome would have served as a skylight and ventilator, and a clock face in the ceiling would have been visible from far below. This clock face registered not only the time but also the wind direction from the weather vane outside. Certainly there was a gallery around the wall at the top of the staircase, for Tycho wrote of having a "free view in all directions" from there. There was another gallery outside "on the top of the house itself," the arches of which are visible just under the cupola on Tycho's elevation drawing.

Between the second floor and the dome itself, probably accessed by the same spiral staircase, was an area Tycho dubbed "the upper story" or "the very top of the house, where round windows [on either side of the entrance tower] are visible [on Tycho's elevation drawing] with eight unheated bedrooms for the collaborators." The warm dining room below was a haven indeed. One of the marvels of Uraniborg was the system Tycho had for communicating with the attic chambers. Cords within the walls ran to small bells in each of the rooms to allow Tycho to summon individual students. Tycho enhanced his image as a magician by demonstrating this system for guests. He would whisper a student's name, and that student would immediately walk through the door as if called by magic through the walls and thin air. Tycho had in fact pulled a cord without his visitor's noticing, then waited until he knew the student must be nearing the door before whispering his name.

Tycho kept a record of the visitors to Hven in his meteorological diary. It was an illustrious procession of nobles and their wives, scholars, and royalty who came to his island and his remarkable home to be wined, dined, entertained, and amused by the communication system and the revolving fountain, to take the air in his gardens, and to compose love poems and heroic verses for one another in Danish and Latin.* Nor was all this accomplished in an atmosphere of sobriety. A letter from Tycho to Bartholomew Schultz— evidently written at the dining table—reported that he and his companions were drinking "one mug after the other, filled to the brim" and exclaimed that such drinking "to the very dregs" was a learned art in itself, though what they were drinking was "not philosophical and least of all theological." "We dedicate these toasts to you to the sound of trumpets, recorders and lutes, and with the sound of sweet song," he told Schultz.

As the years passed, the interior of the house became increasingly elegant in decorative detail and furnishings, and the gardens sprouted

*Tycho's poem "Urania Titani" is widely considered the finest poem any Dane has written in Latin.

not only exotic trees and plants but also aviaries and gazebos. Tycho's goal was to create the Renaissance ideal of Eden, but unlike many of his noble peers, who also cultivated imported fruit trees, herbs, fresh-water fish in ponds, game birds, and animals, he was more than a collector for collection's sake. He was genuinely interested in the study of plants, animals, and birds.

Beyond the earthwork wall an odd collection of facilities sprang up—a paper mill, an instrument works, and later a partly subterranean observatory. The Uraniborg complex, supported by the rest of Hven and Tycho's other fiefdoms in Denmark and Norway, was like no other castle or stronghold of its time. Nevertheless, it fulfilled admirably the concept of manorial self-sufficiency, with Tycho's definition of it including much that the majority of European aristocracy would not have thought essential.

Marring this halcyon vision was the continued unhappiness of the villagers of Tuna. The king's reply to Tycho's complaint in 1578 had been a defeat for them. They were well informed enough, however, to know that they also could appeal to the king. In the summer of 1580 they did, charging Tycho with demanding "harmful, uncustomary burdens of boon work, cartage, and other labor." The family's move into the house required much heavy lifting, and with land under cultivation for Tycho in addition to their own fields, the peasants were doing his plowing, sowing, reaping, and threshing as well as their own. The heating and chemical ovens were insatiable consumers of the wood that had to be hauled across the sound.

Such complaints were not ignored. Tycho made a formal reply. A commission visited Hven and then took more than a year to study its findings and reach a decision. The outcome was an even worse one for the villagers. A new charter issued by the king and his ministers laid down in precise detail what their obligations were to Tycho and what his were to them. It established once and for all that the peasants of Hven were crown tenants, not freeholders. Tradition and possession without written charters and deeds had proved valueless. The new charter decreed that the villagers even had to pay Tycho a fee to graze their swine on the acorns in Tycho's forests in Skåne or else give him the next-to-best swine when they brought the herd back to Hven. Such a fee was customary elsewhere, but no one had ever collected it from these islanders.

Tycho's attitude toward his peasants and his treatment of them were not unusual. There were exceptions, of course, such as his uncle Steen, who reportedly acted toward his tenants at Herrevad "like a mild father" and "did not lay new burdens upon them," which many of them "acknowledge to this day and bemoan with tears that they do miss their good master." However, the charter of Hven served as a prototype for other similar documents all over the kingdom. It did not seem hypocritical for Tycho, in the context of the times, to talk of Uraniborg and Hven constituting a glorious link between humankind and a love whose force drove the universe, nor did he deny that his peasants were part of that humankind. They simply had to fulfill their obligations.

Though his tenants had come to regard Tycho as a monster, and some rival scholars such as Jørgen Dybvad must have had a scarcely better opinion of him, most of Tycho's acquaintances during this period of his life probably found him to be an amiable, charming man. He had many of the unconscious attitudes and assumptions that went with his birth and breeding, but he was not a social snob. He had chosen a commoner for his life partner, and he preferred the company of those who belonged to a scholarly class whose social position was considered beneath his own. However, with the completion of Uraniborg, mounting wealth, close association with the king, and particularly with the remarkable lack of opposition he was meeting as he pursued his dreams, he was undeniably developing an increasingly elitist attitude and an inflated ego that were all the more problematical because they were in many respects well founded.

With success following on success, Tycho was beginning to equate "good" with "what contributes to the splendid work I am doing."

Students, assistants, and peasants should feel privileged to contribute to this magnificent venture that seemed to be favored by God and providence. There were those who did indeed see working with Tycho as likely to be their own closest brush with greatness. Tycho did not stop being an amiable, charming, even considerate man to those who saw things in this light, whose plans supported his own lofty ambitions, or whose intellects were in sync with his. He was also subjugating *himself* to this great cause of reforming astronomy. He could not think it unreasonable to expect others to burn willingly on the same altar, be grateful for the opportunity, and be respectful of the unique responsibility that was his.

The cause to which Tycho was dedicating himself and his great building project, and everyone around him, in the late 1570s was the achievement of a standard of observational precision undreamed of by his predecessors or contemporaries. Even the finest instrument makers had never been asked to produce instruments like those he wanted. At Herrevad he had recognized that it was essential for him to have his own shop where craftsmen who were already skilled could become specialists and devote all their time to his needs, under his close supervision. That had not materialized there. At Uraniborg, he was making sure it did.

Tycho's "workshop for the artisans," as he called his instrument shop, was admirably well equipped, beyond any other such shop in existence, with "mills driven partly by horses and partly by water power." Instrument building was costly in material and labor, with some instruments taking six trained people three years to manufacture while peasant labor kept the mills running.

Tycho kept meticulous records of the manufacture and use of his instruments. He had learned a lesson from the controversy about the nova: Not only did he need to achieve instrumental accuracy to his own satisfaction, but he also had to be able to prove to others that he had achieved such precision. To that end, records had to show which instrument he used for each observation and be cross-referenced to identify all observations made with each instrument. For cross-checking, related observations had to be made with different instruments.

All this recording and documenting, as well as the everyday conversation at Uraniborg in the observatories and at the dinner table, used a vocabulary of astronomy that was elementary to Tycho and his associates and students, and much of which is, in fact, utilized by modern astronomers. Ancient and medieval astronomers, and most in Tycho's time, thought of the stars as fixed onto an invisible "celestial sphere" centered on Earth. (That term is no longer taken literally, but the concept is still used.) The dome of sky visible at night by an observer on Earth is half the complete celestial sphere; the other half is hidden below the horizon. The stars, fixed points on the celestial sphere, serve as reference points for tracking the motion of the Sun, Moon, and planets as they move across this background.

Observing the sky long enough reveals that the stars are not really "fixed." Earth is turning on its axis, so the celestial sphere appears to rotate. Stars rise in the east and set in the west. Their positions in relation to one another do not change while an observer stands and watches them at night, but their positions in relation to that observer and the horizon do change.

The celestial sphere, like Earth, has an equator—the celestial equator, which is on a plane with Earth's equator. Also like Earth it has an axis of rotation that can be thought of as an extension of Earth's axis of rotation. Just as Earth's axis of rotation runs from its North Pole to its South Pole, the celestial sphere's axis of rotation runs from the north celestial pole to the south celestial pole.

The horizon is another great circle (see figure 7.3) that, like the celestial equator, divides the celestial sphere in half. However, while the celestial equator (like Earth's equator) is the same for observers in different places, the horizon is not. For Tycho, on Hven, the horizon was different from what it would have been had he set up his observatory in Basel. Stars whose paths dipped below the horizon at Hven would not have done so in Basel. Thus for any observer on Earth who is not

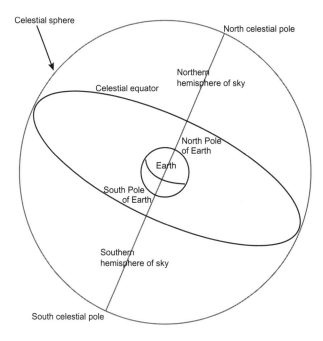

Figure 7.2: On Earth, the equator is midway between the North and South Poles. On the celestial sphere, the celestial equator is midway between the north and south celestial poles, on a plane with Earth's equator.

standing on Earth's North or South Pole, the horizon is tipped in relation to the celestial equator. How much tipped depends on where the observer is located. The celestial equator and the horizon meet at two points, one east and one west of the observer. They are like two hoops hinged together (see figure 7.4).

The zenith is the point directly above an observer's head, regardless of what happens to be up there at the moment or where the observer happens to be standing on the face of Earth. Hence, as is the case with the horizon, an observer can think of the zenith as belonging to him or her personally, while the celestial poles and the celestial equator do not—they are public property, worldwide.

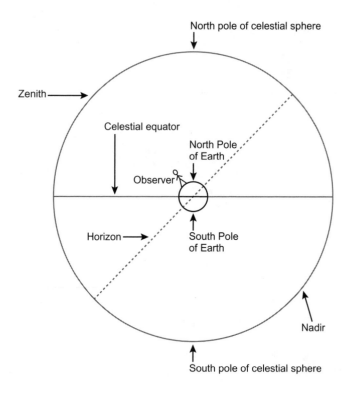

Figure 7.3: Anywhere an observer stands on Earth's face, except on its North or South Pole, that observer's horizon is tipped in relation to the equator and the celestial equator. In more technical language: The observer's horizon is inclined at an angle to the celestial equator.

There is one more hoop hinged into this arrangement (see figure 7.5): Tycho and most of his contemporaries believed that the Sun orbited Earth, completing one orbit in one year, moving in a great circle around Earth. (In fact, when the discussion involves only Earth and the Sun, there is no way of deciding which is orbiting which. The two arrangements are geometrically equivalent.) Astronomy's name for that circle was and still is the ecliptic. The ecliptic is not the same as

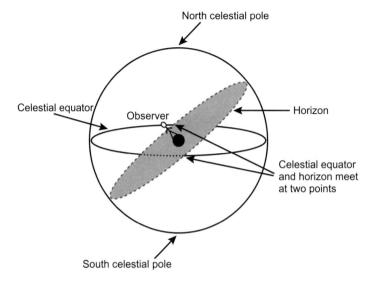

North celestial pole

Celestial equator

Observer

Horizon

Celestial equator
and horizon meet
at two points

South celestial pole

Figure 7.4: Two "hoops," the horizon and the celestial equator, are hinged together and tipped in relation to one another.

the celestial equator, because Earth's axis of rotation (the line drawn
between the North and South Poles) does not run at a ninety-degree
angle to Earth's orbit. Like a fishing bob tilting in relation to the surface
of the water, Earth is tilted in relation to the plane of its orbit. During
the summer in the northern hemisphere, the north pole tilts toward the
Sun. When it is winter in the northern hemisphere, the North Pole tilts
away from the Sun, while the South Pole tilts toward the Sun.

The planets and the Moon also take part in the apparent daily rotation of the sky due to the rotation of Earth, but they also have additional movement of their own. To an observer, most of the time, each
of them appears to move in a great circle around Earth. These orbits
are not the same as the horizon, the celestial equator, *or* the ecliptic.
They are inclined at angles to these and to one another. However, the
angle between the orbit of a planet and the ecliptic is never large.
Pluto's orbit is much more inclined than the others, but Tycho and his

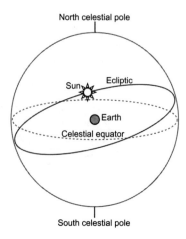

Figure 7.5: For a Ptolemaic astronomer, the ecliptic was the great circular path along which the Sun appears to travel as it orbits Earth. It is another "hoop," with its plane tilted at an angle to the celestial equator.

contemporaries knew nothing of Pluto. They did know that the Moon and planets never strayed outside the zodiac, a belt of sky about ten degrees wide, centered on the ecliptic (figure 7.6). The position of a planet could be pinpointed by saying where it was in relation to the background stars in that zodiac belt: in other worlds, where a straight line of sight drawn from Earth through the planet would end in the zodiac.

(Other terms that are useful for understanding the descriptions of Tycho's instruments are defined in appendix 2.)

It was possible for Tycho and his contemporary astronomers to calculate from one set of measurements to another. For example, knowing the position of a star with reference to the horizon, they could calculate its position with reference to the celestial equator, or to the ecliptic, and vice versa. To make such transformations, they often chose to use a shortcut, an instrument called an armillary—an arrangement of rings showing the relative positions of these circles on the celestial sphere. Tycho spoke of armillaries disparagingly as

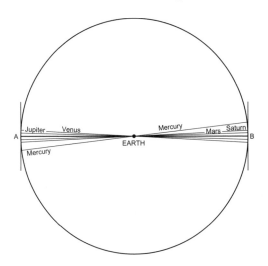

Figure 7.6: The angle between the orbit of a planet and the plane of the ecliptic (the straight line from *A* to *B* in this drawing) is never large, so the planets
are always found near the ecliptic, within the band known as the zodiac, the
width of which is marked with lines at *A* and *B*.

devices for "people who shun labor," but he built and used several of
them himself. One of the wonders of Tycho's island that guests
would later gaze at with awe was an armillary larger than any other
that has ever existed.

The last instrument Tycho commissioned before he began producing them himself at Uraniborg was his *quadrans mediocris orichalcicus
azimuthalis,* or "medium-size azimuth quadrant of brass." It was one of
his favorites in a lifetime of instrument production and a landmark in
the evolution of his instruments. The procedure for using a quadrant
(see figure 7.7) was straightforward enough, but the degree of accuracy
Tycho wanted created problems. The study of the movements of the
planets, as well as positions of such phenomena as comets and novas,
required more than precise viewing of the object in question. It also required a background catalog of fundamental star positions to serve as
reference points. Compiling such a catalog (for he did not find anyone

else's nearly dependable enough for his needs) was one of the most essential tasks of Tycho's career, and it continued for many years. The improvements he made in the development of the *quadrans mediocris orichalcicus azimuthalis* represented significant breakthroughs that were essential to what he was trying to achieve.

The age-old method was to sight through pinholes. Few astronomers before Tycho had demanded enough precision to be annoyed by the deficiencies of this method, much less to do anything about them. However, Tycho found that if the holes were large enough to see through and find the star, the sighting was not precise. The star would not necessarily be centered exactly in the holes, and the position could be off by a fraction. Thus, "driven by necessity" to seek an improvement, he came up with a better *alidade* (the straight piece of an observing instrument that connected the nearer and farther sights). It was such a rousing success that he included a drawing of it (figure 7.8a) much later in his book *Astronomiae Instauratae Mechanica*.

With this new alidade set on its sides, it was possible for Tycho to raise or lower it until he could see the star through slits he marked A–D on his drawing, lined up precisely with the side H–E at the other end of the alidade, while *at the same moment* lining it up so that just as much of the star could be seen through the slit B–C, sighting on the line F–G. In that way he measured the altitude of a star (its distance above the horizon). At the same moment he could look through the slit C–D toward the side G–H, and simultaneously through the slit B–A toward the side F–E. That gave him the azimuth measurement (distance from the meridian; see appendix 2). To study the Sun, he could adjust the instrument so that the Sun's rays shone through the round hole in the far sight and filled a circle drawn on the inner side of the clover sight (not visible on his drawing).

A further innovation that Tycho came up with to improve accuracy of sighting was the use of a cylinder as the more distant sight. Later, for his great mural quadrant, he positioned the cylinder in a rectangular opening in a wall of his house.

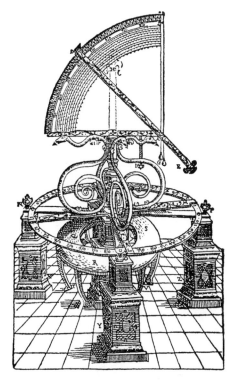

Figure 7.7: Tycho's "*quadrans mediocris orichalcicus azimuthalis*" or "medium-size azimuth quadrant of brass," in a drawing from *Astronomiae Instauratae Mechanica.* To use this quadrant, Tycho positioned it so that its plumb lines— *G* in the picture—showed that one of its straight edges was precisely horizontal and the other precisely vertical, pointing straight up toward the zenith. He rotated the quadrant on its pivot so that the curved edge, or arc, passed through the star or planet whose position he wanted to measure. The azimuth of the star or planet (its distance from the meridian) could then be read off the 360-degree circle within which the quadrant rotated. The sighting arm, a straight piece called the *alidade,* was attached at the point of the quadrant that corresponds to the center of a pie. Like the hand of a clock, its other end was able to move freely along the arc. Tycho and his assistants raised or lowered that end (*D*) of the alidade until, sighting along it from the other end (*E*), they had it pointed at the star or planet. The arc was marked off like a ruler into the ninety degrees represented by this segment of a complete circle, allowing one to measure the altitude of the star (its distance in degrees above the horizon).

Figure 7.8
a.) Tycho's drawing of his new alidade, in *Astronomiae Instauratae Mechanica*. The clover-shaped end (letters A, B, C, D) was the end of the alidade nearest the observer. The square end (*F, G, H, E*) was at its far end, the end that could swing freely along the arc. The clover had slits on four sides, forming a square that exactly corresponded to the square at the other end of the alidade. The width of the slits was adjustable. "By turning one single screw, that is by one single manipulation," Tycho wrote, "it is possible to widen or narrow all the slits simultaneously without any trouble or waste of time."

b.) Using a cylinder as the more distant sight: The diameter of the cylinder was the same as the distance between two slits in the near sight. Tycho lined up the sights so that the star appeared equally bright on both sides of the cylinder when he moved his eye from one slit to the other.

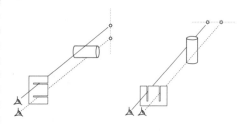

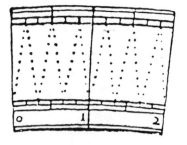

c.) Transversal points: The zigzag lines of dots on the arc of an instrument made it possible to fine-tune adjustments so as to have the line of sight passing through one of these points, allowing much more precise measurements.

Tycho made one final modification to his *quadrans mediocris orichalcicus azimuthalis,* at last fully utilizing the transversal points he had learned about years before when working with his cross staff. In his drawing of the quadrant (figure 7.7), the zigzag pattern of the dots that enabled him to make much finer measurements was visible on the curved edge.

Having his own instrument shop at Uraniborg proved to be an enormous advantage. Not only was Tycho able to supervise manufacture closely; he could also evaluate each instrument after it came out of the shop by using it for observation, studying its quality, identifying its problems, and experimenting with innovative ways of solving them. He could easily return the instrument to the shop any number of times to make the necessary adjustments and corrections or rebuild it completely. Instruments he had only been able to dream of at Herrevad were about to become a reality at Uraniborg.

8

ADELBERG, MAULBRONN, URANIBORG

1580–1588

LIFE AT THE SEMINARIES in Adelberg, where Johannes Kepler, just barely in his teens, matriculated in 1584, and Maulbronn, where he went two years later, was severely regimented. The school day began at four A.M. in summer, five in winter, when all the students, dressed in identical sleeveless knee-length coats, gathered for psalm singing. Every hour had its assigned work, with no free time. All conversation continued to be in Latin, but at this level there was instruction in Greek as well, and also in rhetoric and music. Johannes and his schoolmates now read the classics and the Bible in both Latin and Greek, thus mastering the classical languages while at the same time assimilating the ideas, faith, and values of Western civilization. The higher seminary introduced them to "spherics" and arithmetic.

Though school may have provided the happiest moments in Kepler's childhood and youth, he was oppressed by a series of real and imagined physical ailments and had difficulties typical of his age group when it came to getting along with fellow students. His long, rambling list of "only those who were hostile over long periods" contains many statements such as "I willingly incurred the hatred of

Kloster Adelberg, where Kepler went to school after he had completed Latin
school, in a cut-out from an old forester's map.

Seiffer because the rest hated him too, and I provoked him although
he had not harmed me. . . . I have often incensed everyone against
me through my own fault . . . at Adelberg it was my treachery [under
strong moral pressure from his instructors, Kepler had acted as an in-
former]; at Maulbronn, it was my defence of Graeter." Kepler was
also the butt of insults because of his father's reputation, but he was
particularly hurt when there was envious talk about him: "Why were
all of them all the time jealous of competence, industry of work,
progress, and success?" Each of his schoolmates at this age probably
could have come up with a litany similar to Kepler's. Kepler was candid enough to write it all down.

Kepler was not the colorless, drab individual that some authors
and historians have made him out to be. He had enthusiasms in
abundance. By his own report he, as a boy, "devoted [himself] whole-
heartedly and energetically to games." In his teens he "had a high

opinion of sense of duty, self-control, and industriousness," but "a man who is really useful has to have not only the power of good judgment but also ardor and passion." Perhaps following that last maxim, "I didn't obey reason until my twenty-sixth year" (the year he wrote these words).

Earnestly but exuberantly religious, Kepler had a vivid and adventurous life of the mind. In his spare time, which was scarce, the teenage Kepler took pleasure in attempting to write original poetry in a variety of meters, imitating the ancient forms. He reveled in jokes and puzzles, loved allegories and riddles, liked to play with anagrams, was pleased with paradoxes. Purely for enjoyment, he tried to improve his memory by learning the longest psalms by heart and attempting to memorize all the examples in one of his grammar books. The way his mind flew quickly among various subjects was a joy to him, rarely a problem. In fact, all his life his writing continued to be full of peculiar and interesting leaps from one train of thought to another, leaving readers attempting to follow him puzzled by his mental track.

Though Kepler may not have chosen to "obey reason," he was a serious student, and he thought and prayed a great deal about the religious controversies and his private reactions to them. At Adelberg some of Kepler's young teachers, fresh from Lutheran university at Tübingen and afire with their newly acquired learning, were particularly eager to refute the doctrine of the Holy Communion espoused by the Calvinists. Kepler stubbornly followed his usual practice of accepting little except what he had worked out for himself after listening carefully to the sermons or arguments, praying, and studying his Bible. He was approaching the awkward conclusion that the correct interpretation of the Bible was exactly the one he was hearing condemned by his instructors and from the pulpit.

Kepler was particularly disturbed about the idea held by some that God damns the heathen who do not believe in Christ. That doctrine he could not accept, nor did he keep quiet about it. He also was so bold as to recommend peace between Lutherans and Calvinists, and claimed also to be "just to the Catholics," setting a course that would have tragic consequences for him later, and that would, in fact, eventually leave him no choice but to appeal to Tycho Brahe for a job.

In October 1587 Kepler became a registered student at the University of Tübingen, but the "Stift," where he was to lodge as a scholarship student supported by the duchy of Württemberg, had no room available for that year. He remained for a third year at Maulbronn, as a "veteran," and ended up taking his examination and completing his B.A. degree by examination the following September, though he had not yet attended a class at the university.

❧

WHILE KEPLER WAS a schoolboy, mastering subjects that he and others thought essential for his future, Tycho, with trial and error and his own superior inventive intellect as his teachers, continued to explore the frontiers of astronomy with a rigor that no one else considered essential. One splendid result of this effort were the great instruments of Hven, unique and unsurpassed among the astronomical instruments that predated the telescope.

The first masterpiece Tycho produced in his own instrument shop was a giant globe that became the centerpiece of Uraniborg's library. The instrument maker Schissler in Augsburg had begun to construct it under Tycho's guidance in 1570, but Tycho had left Augsburg to join his dying father in Denmark before it was completed, and he did not see it until he returned five years later. By that time the wood had warped, and there were splits between the pieces.

Nevertheless, Tycho had not forgotten the globe languishing in Augsburg, and in August 1576 he had the poor relic brought to Hven, where his artisans filled in the cracks and restored its shape "by inserting many hundred pieces of parchment." Tycho allowed the globe to sit two years longer while he watched for seasonal changes in its wooden structure. Finally in 1578, satisfied that "it stayed completely spherical at every point," he had it surfaced with brass sheets "with such great care and accuracy that one might believe the globe to be of solid brass." After waiting another year to find out whether the globe would still stay completely spherical, he had the equator and the zodiac etched onto the brass "and divided each degree of these circles accurately into sixty minutes of arc by means of transversal points according to our custom." By the time he had finished, his remark that he had the globe made "at no small cost" was an understatement.

All this effort and expense were not merely to produce a decorative piece. The globe represented the celestial sphere and allowed one to view that sphere from the "outside." Transforming trigonometrical coordinates, which was necessary if an astronomer knew the altitude and azimuth of a celestial object and wanted to calculate from them its declination and right ascension (see appendix 2), was a tedious undertaking in Tycho's day. The globe made this process considerably easier.

Tycho later included a drawing of this remarkable instrument (see color plate section) in his book *Astronomiae Instauratae Mechanica,* which shows the globe girdled by a platformlike ring resembling a ring of Saturn. The ring represented the horizon for an observer at Uraniborg. The circle seen outlining the perimeter of the globe was the meridian (see appendix 2). The globe rotated on an axis running from the north celestial pole (I) to the south celestial pole (K). There was another rulerlike strip (somewhat right of center in the drawing) running from the zenith (B) to the horizon, which was the equivalent of the curved edge of a quadrant. It allowed Tycho to measure altitude on the globe. Fixed at the zenith, it could be moved around the horizon at its other end to measure azimuth. Two other lines on the globe that are visible in the drawing were the equator (farthest to the left as it reaches near the top of the globe) and the ecliptic. The support for the globe, from the ground to the horizon ring, was about five feet high, and the globe itself measured almost six feet in diameter.

Tycho proudly referred to his great globe as "a huge and splendid piece of work" and wrote that "a globe of this size, so solidly and finely worked, and correct in every respect, has never I think been constructed up to now . . . anywhere in the world. (May I be forgiven if I boast.)" It became the chief conversation piece of Uraniborg and the envy of visitors. He entered on it the positions of all the stars he had observed and cataloged—as they would appear in the year 1600. His goal was a thousand stars, "so that all the stars that are just visible to the eye were entered on the globe." Defending the length of time it took him and his artisans to finish the globe and for him to catalog the stars and enter them on it—the task would eventually require about twenty-five years—Tycho used words that many have taken as a motto for all his work: "If it has been done well enough, it has been done quickly enough."

In the next two and a half years after the completion of the globe in 1580, Tycho and his artisans on Hven produced more large instruments, most notably two quadrants, inaugurated in 1581 and 1582. Tycho named the first his "large quadrant" and henceforth referred to the old *quadrans mediocris orichalcicus azimuthalis* as his "small quadrant." The second was his "great mural quadrant," an artistic as well as scientific masterpiece that more than any of his other instruments has come to symbolize Tycho Brahe.

Tycho built the great mural quadrant into the structure of his house, using a section of wall constructed along an astronomically precise north-south line. The instrument, a solid brass arc six and a half feet in radius, five inches wide, and two inches thick, was mounted on the wall. On this quadrant the curved edge of the pie slice was *nearer* the observer. Movable sights were clamped onto the arc. The engraving Tycho later included in *Astronomiae Instauratae Mechanica* (see color plate section) shows them set near twenty and seventy-five, with one of Tycho's assistants peering through the one near twenty. There was no physical connection (such as the alidade was in earlier quadrants) between these near sights and the farther sight—the cylinder in the opening in the wall visible in the upper left of the engraving.

Before 1587, the wall on which the quadrant was mounted probably remained blank. Then Tycho commissioned Steenwinkel, who had
helped design and build Uraniborg, to paint scenes symbolizing
Tycho's palace and his work, framed by six arches. Steenwinkel's painting showed the basement with its alchemical laboratory and furnaces,
the library on the floor above it with the great globe, and the observatory above that. Hans Knieper, the finest landscape artist in Denmark,
was working at Elsinore, and Tycho brought him to Hven to paint a
distant landscape for the background, visible through the arches of the
observatory level and above them. For the life-size portrait of himself,
seated in the foreground within the arc with a dog at his feet, raising
his arm to point at the front sight of the quadrant, Tycho commissioned Tobias Gemperle, a painter he had met during his 1575
European sojourn and brought to the attention of King Frederick.
Frederick had named him court artist, and Tycho had previously commissioned him to paint the altar of St. Ibb's Church. Tycho was extremely pleased with the portrait in the mural quadrant. "The likeness
could hardly be more striking," he wrote, "and the height and stature
of the body is rendered very realistically." He regarded the background
mural as the artistic masterpiece of Uraniborg.*

Tycho was also experimenting with *armillaries,* instruments consisting of arrangements of rings showing the relative positions of the
various circles on the celestial sphere such as the meridian, the celestial equator, and the ecliptic. Tycho planned to do extensive work on
the planets, and armillaries were particularly useful for calculating
the coordinates involved in planetary observations. He had begun
work in 1577 on a small model with only three inner rings.

*The engraving also shows a man in the lower right-hand corner noting the time of the obser-
vation, while another opposite him transcribes the observation into a log book. These figures
were not part of the mural itself. They were part of the engraving of the mural as it appeared
in Tycho's book.

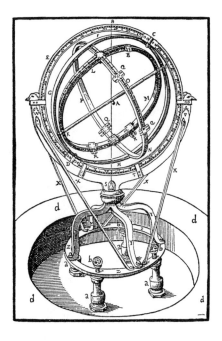

Figure 8.1: Tycho's first armillary. Drawing from *Astronomiae Instauratae Mechanica.* The outermost ring represented the meridian (see appendix 2). The devices marked *C* and *D* represented the north and south celestial poles and are the points on the meridian ring on which the next smaller ring is attached. *C* and *D* could be raised or lowered along the meridian ring until they corresponded with the latitude of the place at which the observer was located, with *B* (from which hangs a plumb line) representing the zenith. The next smaller ring served to carry a slightly smaller ring representing the ecliptic, which carried a fourth ring for measuring latitude.

IN THE EARLY 1580S, having spent considerable money, time, and ingenuity on producing by far the best instruments in existence, Tycho still knew he lacked the tools necessary for what he hoped to accomplish. As a case in point, his armillary tended to bend and flex unpredictably as the rings were adjusted to different positions. A smaller, lighter armillary would not give him fine enough readings, while a larger one would bend and flex even more. Tycho set his mind to solving these particular problems, and this time, as he had done when he designed his mural quadrant, he thought big and, at the same time, simple. He came rapidly to the conclusion that in order to accommodate a successful design he would need a ground-level observatory beyond Uraniborg's perimeter wall.

By 1583, Uraniborg was too crowded. Tycho and his assistants were stumbling over one another. Instruments already in use or in various stages of construction threatened to overflow the space available, and Tycho had not by any means stopped planning new instruments. The large ones he now had in mind would work much better with access to all 360 degrees of the sky, which neither Uraniborg's large observatories nor the smaller satellites provided. Outside the perimeter wall, with some excavation, he could have the advantage of being able to build an amphitheaterlike structure around each instrument, allowing an observer to position himself on a level high above the base of the instrument. The shelter of the amphitheater would prevent gusts of winter winds from affecting sensitive readings, not to mention chilling the observer. Tycho also saw the auxiliary observatory as a way of separating his assistants and establishing better control over the accuracy of their findings. Some would continue to make observations from the castle, others from these new "cellars." They would not get in one another's way and would also not compare results and make adjustments to them before he had a chance to study findings that disagreed and think about the implications.

There was a small rise in the landscape not far beyond the south corner of Uraniborg's outer wall. Tycho decided that a structure there would not spoil the symmetry of the house or the gardens, and the wall and the house would block only an insignificant low portion of the northern heavens, the least interesting direction. He set the islanders digging again, and he constructed "with no small difficulty and expenditure, a subterranean observatory." He christened it Stjerneborg, "Star Castle."

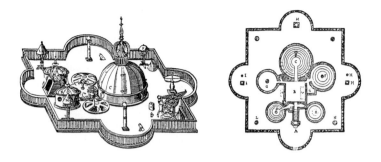

Figure 8.2: Tycho's drawings of Stjerneborg, the partially subterranean observatory that he built outside the perimeter wall of Uraniborg, from *Astronomiae Instauratae Mechanica*.

Stjerneborg (see color plate section) was designed with five great cellars or amphitheaters to house a giant armillary that figured largely in Tycho's observational plans, a revolving quadrant, a zodiacal armillary, a large steel quadrant, and a four-cubit sextant. Each cellar had a roof that could be removed or swung aside. There was ample storage for other instruments as well, and space to use them. Tycho's design also did not neglect his own comfort or that of his assistants. There was a "heating installation," a bed for Tycho "when accidentally there were clouds and we could not enjoy a constant clearness of the sky," and a second larger bed to be shared by others.

Despite being purpose-built as an observatory that pushed the boundaries of what such a building should be and what it should allow its users to achieve, Stjerneborg was far from strictly functional in design and decor. Tycho and Steenwinkel drew plans with the same attention to symmetry, harmony, and detail that characterized Uraniborg itself. A significant difference was that Tycho was by this time much more preoccupied with his self-image as a man of stature and wealth, with classical roots, the equal of kings, the greatest of all living astronomers, occupying a preeminent place in history. Stjerneborg was laden with symbolism to convey this image. Above the entrance stood three elaborately carved lions with crowns on their heads. On the back of the portal an inscription in gold letters sang the praises of Tycho and his instruments. Beyond the entrance and several steps down from it was the warming room that gave access to the five round cellars, and this and other subterranean rooms were embellished with poems inscribed in gold letters. The warming room walls displayed seven portraits of astronomers from the ancients to Tycho, with an eighth portrait of a future astronomer named "Tychonides." The message was clear: Tycho was the equal of the greatest astronomers in history, and he anticipated that Tychonides would come from his own lineage.

Tycho was diverting some of the attention previously reserved for

Figure 8.3: The great equatorial armillary as illustrated in *Astronomiae Instauratae Mechanica*. To find the position of a star, an observer stood on a tier of the amphitheater, behind the half circle (O) representing the celestial equator. Through the movable sight (R), which could be positioned anywhere along this half circle, he peered toward the axis pole (B) and moved the sight along the half circle until, looking through the slits of the sight, he saw the star on both sides of the axis pole. Rulerlike markings on the half circle indicated, from the new position of the sight, the right ascension of the star (distance in degrees east of the prescribed meridian established by the position of the Sun at the vernal equinox). In order to compare this finding with the right ascension of another star whose position was already known, two observers sighted from the half circle, one for each star. They learned the difference between the stars' right ascensions by noting the distance on the arc between the new positions of the sights.

To find a star's declination (distance above the celestial equator), assistants pivoted the armillary on its axis until the large complete ring had one edge toward the observer and the other toward the star. There were two alidades— the "fan blades" that met in the center at a cylinder (E) with their other ends on the large ring. The drawing shows those ends at two positions on the ring, (F). Like hands on a clock, the alidades were fixed at the center and moved along the ring. Assistants moved one until, sighting along the alidade, the observer saw the star on both sides of the central cylinder (E). The new position of the sight indicated the declination of the star.

To double-check an observation, Tycho gave the entire apparatus a half turn on its axis and used the other alidade for the same measurement. As he informed his readers, "the two values found should agree with each other."

astronomy to promoting that image. He was eager for the stream of
scholars, intellectuals, and highborn curiosity seekers, who regularly
detoured in their travels to visit Hven, to recognize his greatness and
also realize that, though his descendants could never be noble
Brahes, the mantle of their father could fall onto their shoulders in a
far more significant way.

Tycho's "great equatorial armillary" (figure 8.3) was destined to
become the most famous of Stjerneborg's instruments. Like the great
mural quadrant inside the house, it was built into the building's
structure. The foundation for its axis was put in place in December
1584, but it was not until the following summer solstice that the
massive instrument was ready for use.

It was difficult to see that the instrument *was* an armillary, for
most of the familiar rings (compare with figure 8.1) seemed to be
missing. Except for brass alidades (the two "fan blades" that ran from
the center at E on Tycho's drawing to the outer edge, marked F) and
graduation strips to act as rulers on the rings, the great equatorial
armillary was made of wood for easier handling and to minimize the
distortions caused by the weight of metal. It pivoted on a hollow steel
axle (the pole marked B in the drawing). At first glance, the entire
apparatus looks to be strangely skewed from plumb, but take a globe
of the world and tilt it so that Denmark is on "top" and the skew
makes sense.

Tycho supported the giant armillary at its lower pivot point with
a half-buried stone pillar (partially visible at the lower left corner of
the picture) topped by a globe supported by a splendidly carved figure of Atlas. A wishbone-shaped stone structure supported the upper
pivot point, and the two legs of the wishbone flanked the door to the
passage to the warming room and the crypt beneath the armillary.

After the completion of the mural quadrant and the great equatorial armillary, Tycho and his shop went on to produce a revolving
wood quadrant and to revamp the old 1581 "large quadrant" by giving it a stronger base and a pivot at the top to stabilize it. It was

henceforth known as the "great steel quadrant." This done, Tycho finally felt that he no longer needed to use his older, less successful instruments to check daily observations. When observations with his newer, larger instruments agreed with one another, he was confident that he had achieved an extremely high degree of accuracy. He had at last made enormous strides toward fulfilling his dream of instruments to create a new astronomy.

9

CONTRIVING IMMORTALITY

1581–1588

IT WAS NOT ONLY in its design and symbolism that Stjerne-borg was a direct bid for immortality. The urgency with which Tycho went about his instrument development and the construction of the new observatory stemmed in large part from a decision he had made as early as 1581 about his future research.

If Copernican astronomy or his own evolving Tychonic system was correct, the planet Mars came closer to Earth than the Sun did. If the venerable Ptolemaic system was correct, Mars never came as close as the Sun. Hence there was a way of deciding the contest between the systems: Find out whether Mars does, in fact, come closer than the Sun. One way of doing so was to measure Mars's parallax and compare it with that of the Sun. However, Mars's parallax had never been measured. No observation had ever been able to show that Mars even *had* a parallax. Tycho took on the challenge, a project that would require all his ingenuity and the finest instruments in the world.

At this point, Tycho had not yet conceived in full his "Tychonic system of the world," but he had come a good way in his thinking about how one might devise a compromise between Copernicus and Ptolemy. Three years after the Copenhagen lecture series, in which he had attempted to retain the essentials of the Ptolemaic system while eliminating the need for an equant, he had been considering the possibility, suggested by others, that Venus and Mercury orbit the Sun, while the Sun and the outer planets orbit the unmoving Earth.

In 1580 a young scholar named Paul Wittich had spent three or four months at Uraniborg and shared his own attempts to solve the same problems. Their conversations were a great stimulus to Tycho. However, although he would later claim to have done so earlier, it probably was not until 1584 that he finally arrived at the full Tychonic system in which *all* the planets orbit the Sun, while the Sun and Moon orbit Earth—an arrangement later embraced by Jesuit scholars who opposed Galileo. This was the beloved intellectual child that Tycho would spend the rest of his life defending, and that he would guard with self-destructive paranoia. Though it retained an unmoving Earth, it was, in fact, the geometric equivalent of the Copernican system. (see figure 9.1)

Tycho had used parallax measurements to try to find the distances to the nova of 1572 and the comet of 1577. In order to use the equivalent of two eyes (as in the finger-before-the-face demonstration) in observing the parallax shift of Mars against the background stars, one "eye" must be very far away from the other. About ninety years after Tycho attempted the measurement, Gian Domenico Cassini succeeded in measuring Mars's parallax by placing one observer in Paris and another in Cayenne in South America. Such an option was not available to Tycho, nor were Cassini's telescopes. However, Tycho had another method, the one he had used for the nova and the comet: An observer could stay in place and let the rotation of the celestial sphere (or of the Earth, if he thought like a Copernican) transport him from one viewing position to the other. A shift in the position of a celestial object viewed in this manner is called *diurnal parallax*.

Tycho knew that the parallax of Mars as viewed from Earth would be tiny and that the attempt to measure it would stretch his capability of precise observation to the maximum. Mars does of course

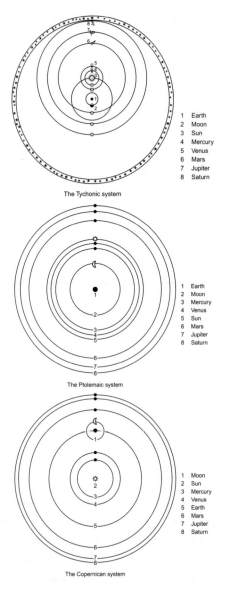

Figure 9.1: The Tychonic system of the world, compared with the Ptolemaic and Copernican systems.

come nearer to Earth than the Sun, but not so near as to make its diurnal parallax ever more than twenty-seven arcseconds. Observing Mars's parallax shift was, in fact, not possible with Tycho's instruments, but he did not know that. Tycho, in fact, thought the Sun was much closer than it actually is, and that its parallax was three arcminutes rather than the nine arc*seconds* we know it to be.

If three arcminutes was the right measurement, and if the Ptolemaic model was correct and Mars was always farther away than the Sun, Mars's parallax would always be *smaller* than three arcminutes. But if Copernicus (or the still evolving Tychonic system) had it right, Mars would come to within half or even a third the distance to the Sun and, when it did, have a noticeably *larger* parallax than three arcminutes.*

The best time to make the measurement was when Mars made its closest approach to Earth, and that was when it was at "opposition"— on the opposite side of Earth from the Sun. Some oppositions bring Mars closer to Earth than do others, and unfortunately, the oppositions when Mars was closest were in summer, when the Danish nights were too short for making the measurement. Nevertheless, Tycho hoped that Mars would approach near enough to Earth at the winter oppositions. If the Copernican or the Tychonic system was correct, he judged that Mars's diurnal parallax at the winter oppositions would be about five arcminutes, a shift that would be just barely possible to observe with his best instruments. On this tiny possibility, he pinned all his hopes.

The great mural quadrant was ready in June 1582, and Tycho began using that magnificent instrument, placed so conveniently across the corridor from his winter dining room, to find the positions of the background reference stars he needed for measuring the parallax. In late December and January, Mars was at opposition. Tycho and his staff made observations of Mars on the eastern hori-

*Recall that the closer one holds a finger to one's eyes, the larger the shift against the background appears to be.

zon in the evening and near the western horizon in the morning. The observations showed no parallax.

So far it was victory for Ptolemy, defeat for Copernicus and Tycho. However, that was not to be the end of the matter. This campaign would involve Tycho for much of the remainder of his life at Uraniborg and be his chief motivation for building more precise and powerful instruments and the new observatory to support them. Searching for an answer that they would never find, he and his assistants, quite unaware, were doing the essential background research for Johannes Kepler.

Tycho had to wait two years, after the 1582–83 observations, for the next opposition of Mars, in January 1585. This time his results were nonsensical—a *negative* parallax. Also, Mars's retrogression— the "backing up" that occurs during opposition—was smaller than expected. Perplexed, he wondered whether the refraction of light by Earth's atmosphere might be to blame.

Refraction is the change in the direction of light waves as they pass from one medium to another, in this case from empty space to the atmosphere. The most familiar demonstration is the way a rod appears to bend when partly immersed in water. Refraction was not well understood in Tycho's day. In 1585, though he had encountered refraction in studies of the Sun, he did not know whether starlight was refracted or, if it was, whether a planet would suffer the same degree of refraction. But it was reasonable to suspect that refraction might be affecting the accuracy of his Mars observations. In the evening, Mars was a few degrees closer to the horizon than the star Tycho was using for comparison, and he wondered whether this might cause Mars to suffer greater refraction than the star. Tycho began to devise observations that would help him determine the degree of refraction to be expected for stars. As he continued to ponder the problem, the instruments of Stjerneborg finally came on-line.

When Mars next came into opposition in March 1587, Uraniborg and Stjerneborg were poised for the assault as never before, with the equatorial armillary and two large azimuth quadrants installed in

Stjerneborg and numerous assistants awaiting Tycho's orders. Shortly before that opposition, Tycho wrote a letter to Landgrave Wilhelm of Kassel, voicing his optimism about finding the parallax.

March 10 was a typical night.* In some cases there were simultaneous observations with the sextant and the quadrant, and sometimes with the equatorial armillary. Most of them were noted in Tycho's handwriting.

6:27 P.M.—a meridian altitude of Jupiter [its altitude above the horizon as it crossed the meridian], made using the revolving quadrant, followed by a series of lunar positions and then a measurement of Mars's declination, all made using the equatorial armillary.

Just before 7 P.M.—measured distance between Mars and the star Regulus, then between Mars and Cauda Leonis, and then between Mars and Arcturus.

For the next hour—triangulation of distances between these stars, then some measurements of Saturn and Jupiter.

8:30—break

9:45—a few observations of Regulus and Spica "for learning the refraction of Mars"; and shortly after that a few positions of the Moon.

11:22–12:02 (twenty minutes before and after the meridian passage of Mars)—assistants recorded distances with the equatorial armillary and trigonal sextant.

Shortly after midnight—bedtime. [Perhaps many of them piled up on the beds in the underground room at the center of Stjerneborg.]

---

*Owen Gingerich and James Voelkel are two modern-day experts on the life and work of Tycho Brahe and the astronomy of his era. They have studied Tycho's journals and laid out in all its paradoxical intricacy the chronology of Tycho's campaign to find Mars's parallax. The reconstruction of the activity that took place on the night of March 10 comes from their article "Tycho Brahe's Copernican Campaign."

4:14–5:16 A.M.—a new set of Mars observations with the armillary and sextant, once more for finding its angular distance from Arcturus and Cauda Leonis (which by this time stood above Mars in the sky) and to Spica. [The Mars to Arcturus measurements showed that refraction was altering Mars's position by an amount that was consistent with Tycho's refraction table for the Sun.]

Night after night in the starlit darkness on Hven near the grazing cattle and sheep . . . the trigonal sextant carefully adjusted against the sky . . . the great equatorial armillary swinging around into place . . . the planet blinking against the sites of the revolving quadrant—from the observatories in the castle, from the amphitheater steps of the sunken observatory beyond the walls—the magnificent barrel-chested, red-bearded astronomer and his retinue of assistants and interested visitors stayed up most of the night and carried out this systematic program of observations . . . all to capture and pin down a planet that had so tantalizingly orbited Earth—or was it the Sun?—since the world began, to fix its positions and understand its motions as no one had ever done before.

Tycho put all this data together and computed what it meant, correcting the position of Mars to take refraction into account, based on his solar refraction table. He was elated with what he found. He had measured a diurnal parallax for Mars of about five and three-quarters arcminutes, meaning that the planet *did* come closer to Earth than the Sun did. The parallax agreed with Tycho's computation of what it should be in the Copernican model. For a little while, Tycho had reason to hope that he would indeed be remembered for one of the greatest achievements in the history of science—overturning Ptolemaic astronomy.

During this same year, 1587, Tycho's book about the comet of 1577, much of which he'd written shortly after the comet's appearance, was finally on press, and Tycho felt he had to add a chapter describing the position of the comet relative to the planets. This was no

small decision, for in doing so he would have to announce his conclusions about the planetary system.

❧

T H E   D E T A I L S   O F the full Tychonic system had begun to come
together in Tycho's mind in about 1583. After that, during numerous
lengthy discussions in the Winter Room, nights of observations and
days of calculation, he refined his model. In 1584, though not yet
completely satisfied with it, he sketched out his precious idea with a
piece of chalk on a green tablecloth for a guest, Erik Lange. And just
there the crack first appeared in Tycho's supreme self-confidence. A
chain of events began that would eventually turn this proud, well-
focused man into someone at times resembling a wounded, cornered
animal. Perhaps the change had started earlier and happened more
gradually, but this is where it first appears in the surviving documents
about him.

Erik Lange was a good friend and a relative by marriage. He had
been present at the dedication of Uraniborg. By 1584 he governed
Bygholm Castle in Jutland, which meant that when he visited he
came with a considerable retinue. Among them, this time, was a
young man named Nicolaus Reimers Bär, a surveyor. Though he was
the son of a pig farmer from the Dithmarsch in Holstein, Bär was a
gifted mathematician and a likely candidate to join Tycho's band of
assistants. He composed some flattering verses in an attempt to ingratiate himself with Tycho.

Tycho and his household sensed something underhanded, even
sinister, about Bär, and Tycho took a distinct disliking to him. The
problem initially had to do with the fact that while Tycho was entertaining Lange and other noble guests, Bär lurked in the library and
leafed through Tycho's manuscripts, making notes on scraps of paper.
He also surreptitiously examined and sketched Tycho's instruments.
One of Tycho's assistants, Anders Viborg, called attention to this peculiar, secretive behavior. Viborg baited Bär, leading him into pre-

posterous arguments, which ended with Bär in a rage. Tycho silenced one of his outbursts at the dinner table with the jocular dismissal, "Those German fellows are all half-cracked." But Tycho was far from taking the matter lightly. Before he sketched his system on the table-cloth, he insisted Bär leave the room, and he erased the sketch before Bär returned.

The investigation of Bär moved to a new stage. His quarters were changed so that he shared a room with Viborg. While Bär slept, Viborg managed to empty one of his pockets and found "four whole handfuls of tracings and writings." When Bär discovered that some of his surreptitiously written notes were missing, he began "shrieking, weeping, and screaming so that he could hardly be calmed down." Both Tycho and Lange promised that anything that actually belonged to him would be returned, and both men, on the surface, treated the episode as a disagreeable joke. But Tycho worried that Bär might have seen material having to do with Tycho's new planetary model.

In the spring of 1586, Bär surfaced at the court of Landgrave Wilhelm IV of Hesse, in Kassel, where Tycho himself had visited with such success eleven years earlier. Tycho was still carrying on a friendly correspondence with the landgrave who had so enthusiastically encouraged him and praised him to King Frederick. Wilhelm now heard from Bär of a new planetary system that Bär claimed to have invented during the past winter. The landgrave was deeply impressed and commissioned a mechanical model of Bär's system.

When Tycho took up his pen in 1587 to write the new material for his book about the comet, he was still unaware that Bär had plagiarized his model. His unease as he remembered Bär's visit was nevertheless sufficient motivation for publishing his planetary system as quickly as possible. He ended his manuscript with great flair, describing his model, much refined during ten years of research on Hven. All the planets revolved around the Sun, while the Sun revolved around Earth. The orbit of Mars intersected the orbit of the Sun, which was quite possible, Tycho insisted, because the orbits were *not* spheres made of crystal. But Tycho's iconoclasm went further than shattering the "crystalline spheres"—about which astronomers had harbored some doubts even as early as Ptolemy. He had concluded that the comet's velocity was irregular as it moved around the Sun, defying another ancient assumption, that celestial motion must be uniform. He even moved beyond Aristotle, Ptolemy, and Copernicus by suggesting that its orbit was egg-shaped, not circular as he had previously thought and as earlier astronomy required all orbits to be.

Tycho finished the book with a critical overview of all other literature about the comet, a feature that would become a standard part of any scholarly monograph but was revolutionary for his time. Tycho's book became a model for future scientific publications and the definitive book about the comet of 1577.

Mysteriously, Tycho said nothing in his book about having observed the diurnal parallax of Mars in 1587.

During the years when Tycho was planning and executing his assault on Mars's parallax, he was also involved in a related campaign to ensure and link the futures of his family and his work at Uraniborg. Tycho was nearing forty. By all outward signs he was still at the height of his physical and mental powers, but in the sixteenth century a man could not expect good health to last much beyond forty. The work he had begun, he felt, was too important to be allowed to expire when he did. Accordingly, the dedication stone of Stjerneborg was inscribed with words prophesying that posterity would preserve this observatory for the advancement of astronomy, the glory of God, and the honor of Denmark. The portrait on the wall inside representing "Tychonides," the future master astronomer who would carry on that work, was a blatant claim that this astronomer would be none other than one of Tycho's children. The symbolism and rhetoric were in place. The practicalities were more problematical. The future of Tycho's sons and daughters and of his observatory were, in fact, precarious. Both Danish law and tradition dictated that because of their lowborn mother, Tycho's children could not inherit the fief of Hven.

Kirsten and Tycho had found ways to adjust to the problems created by her status. She was mistress of the household at Uraniborg and probably enjoyed as much respect and deference from Tycho's assistants and servants as any noble lady would have. However, when Tycho attended the weddings, christenings, and funerals of aristocrats, he always went without her. When noble and royal visitors, such as Queen Sophie herself, and James VI of Scotland, visited Uraniborg, Tycho's sister Sophie served as hostess for the splendid banquets and festivities. Tycho's relatives and Kirsten's never mingled.

In June 1580 there had been an ominous new royal ordinance condemning common-law marriages as "an evil, scandalous life with mistresses and loose women, whom [men] keep in their houses and with whom they openly associate, brazenly and completely without shame, just as if they were their good wives." The ordinance demanded that the clergy separate such couples and, if the couple resisted, deny them the sacraments and rites of the church.

Since all over Denmark pastors of parish churches owed their jobs to the lords of their manors, it was difficult to enforce the ordinance among the nobility. Tycho did stop going to Communion—perhaps so as to deny the church the opportunity of banning him, or to make things less awkward for the pastor of St. Ibb's. But he and Kirsten went on living as before. Other noblemen married their non-noble "wives," and this provoked yet another royal ordinance, in June 1582, which reinforced the ban on children of such marriages inheriting nobility, land, estate, coat of arms, or family name. The same ordinance, however, made it clear that a father could give money and personal property to these children while he was still alive, which they could keep on his death.

Even before the two royal ordinances, Tycho had begun to contrive a way to link the future of Uraniborg and the future of his children. Kirsten had given birth to a son at Uraniborg in 1581, and they had named him Tycho. A daughter, Cecilie, was born in 1582, and a second son, Georg, in 1583. In 1584 Tycho's old preceptor, Anders Vedel, came for a visit, and the two men wrote a draft for a royal patent granting the island to Tycho and his male issue, provided they use it and its facilities for the pursuit of mathematical studies. To grant a fief in perpetuity was rare, to grant it to commoners was unheard of, but it was not unheard of for commoners to hold the position of university professor or head of a secularized monastery with an income derived from a landed benefice, in some ways the equivalent of a fief. Tycho and Vedel, in drafting the patent, implied that Uraniborg had more in common with a university than with a traditional fief, and that the directorship of Uraniborg was like a professorship or headship.

Tycho and Vedel had judged the situation well. When Tycho presented the proposal to Frederick, the king readily approved it, with the queen as witness. Unfortunately, nothing was written down, and no actual patent was issued.

King Frederick's ill health had been a source of concern for more than a year, and he died on April 4, 1588, not long after he had given verbal approval to Tycho's proposal. Frederick's son Christian was still a child, and a regency council assumed the government of Denmark. In spite of the inevitable atmosphere of upheaval, Tycho had no reason for concern, for the new government was packed with his friends, relatives, and allies. In August 1588 he presented his plan for the future of Hven to the Regency Council, which not only issued the patent with a glowing statement of its desire to perpetuate the astronomical work on Hven far into the future, but also endowed Uraniborg with ecclesiastical incomes from canonries and other church offices, implying it could be headed by a commoner. Best of all, the patent laid down an order of succession for Uraniborg, giving preference to Tycho's sons or sons-in-law. It referred to these descendants as "Tycho Brahe's own"—the only official recognition that Tycho had children.

As favorable as this outcome was, Tycho did not allow matters to rest there. To avoid any confusion either in that time of political turmoil or later when the young king came of age, he obtained a patent signed by the entire Rigsraad and the Regency Council. He also persuaded Queen Sophie to put in writing that she could remember her late husband Frederick II stating his intention that one of Tycho Brahe's own children would become head of the observatory. It seemed that Uraniborg would, as the inscription at Stjerneborg prophesied and Tycho had hoped for so long, become a permanent research institution under the directorship of his heirs.

10

THE UNDERMINING OF HUMAN ENDEAVOR

1589—1591

I N 1 5 8 9 Tycho Brahe was at the peak of his career, renowned in scholarly circles throughout Europe and approaching his forty-third birthday. Johannes Kepler was seventeen years old and waiting for an opening at the Stift in Tübingen. Finally, in September, space was available. Already in possession of a baccalaureate degree, he set off for university. Traveling through the forests of the Schönbuch, he carried with him only books and a few personal possessions, a stark contrast to the accoutrements that a wealthy young man like Tycho would have taken along.

The castle of Hohentübingen sat like a mother hen over the university town that huddled beneath it in the valley of the Neckar River. Narrow streets with closely packed high-gabled houses led from the riverbanks to the foot of the castle promontory. Kepler threaded his way through these streets and found the Stift, where he would study and have his lodging. The buildings were old, for this had been an Augustinian monastery before the Reformation. In Kepler's time it was a seminary for scholars who were "children of poor, pious people, with an industrious, Christian and God-fearing character." Somehow the question of Kepler's father's piety and in-

The Stift in Tübingen, where Kepler lived and studied during his university years.

dustry had been overlooked, and Kepler had been accepted. His instruction, room, and board were free, and he had a scholarship of six gulden annually for other expenses. Katharina's father had placed at his grandson's disposal the yield of one meadow "for better and more dignified upbringing." Thus Kepler was well provided for as he moved into the Stift with other young men in their teens from all over Swabia who, like him, aspired to careers of service to the duke or the church. In his second year, on the recommendation of the magistrate of his native city, Weil der Stadt, Kepler received a further stipend of twenty gulden. There were few periods in his life when he was so free of financial worries as during these university years.

Tübingen, like the University of Copenhagen, was steeped in the Philippist philosophy of university and seminary teaching. Though education at the Stift led to a specific goal and allowed students few choices about what they would learn, it was not a narrowly focused trade school. Theological studies didn't even begin until the third year. Before that, in the Philippist tradition of broad education, Kepler had to complete two years of ethics, dialectics, rhetoric,

Greek, Hebrew, astronomy, and physics. An exam in the spring of the second year marked the end of these studies in the arts faculty. After that came two or three years of theological work. Kepler was closely supervised and received grades every quarter. The Stift regulated student behavior almost as rigidly as the lower schools he had attended, and it expected its candidates in theology to avoid the disorderly student life enjoyed by others in the university.

Johannes Kepler was in his element amid all this knowledge to be had for the taking. Rarely has a young man been better equipped to make the most of an opportunity. Soon he had a reputation with teachers and students for being diligent, sedate, and pious, and also for being good at casting horoscopes—a highly valued skill, as Tycho Brahe's experience attested. According to Kepler's own report, he managed to avoid conspicuous shortcomings except for a few outbursts of temper and a thoughtless prank or two, but he still had problems getting along with some of his fellow students. Particularly, he disliked one young man named Kölin, who wanted to be his friend. "Although [Kölin] once made friends with me he always argued with me," wrote Kepler, and complained that an argument with Kölin was more like a "lovers' spat," though most of these arguments seem to have been about work. "With nobody else did I have a sharper and longer competition," Kepler wrote.

Kepler's work habits, though they were certainly productive, were (and would continue to be all his life) a source of some disquiet for him. He was in a state of "permanent repentance about lost time and permanent loss of time through my own fault." He also admitted, "Although I am very industrious, I am the harshest hater of work. But I work for my thirst of knowledge. I am never lacking an object of my desire, my burning eagerness, to do research on difficult matters." His enthusiasms often went beyond his capacity to carry through on them. "In my eagerness, I talked myself into a lot of things that looked easy, but that were difficult and time-consuming in the carrying-out, because the mind is finer, faster, and quicker

than the hand." His mind continued to leap quickly from one matter to another, sometimes among apparently unrelated subjects. "I
talk well and I write well, as long as nothing is pushing me except
what I have already thought of, but in reading and writing I continually start thinking about new things, words, figures of speech, arguments, new insights and understanding, what should be said and
what should not be said."

Among the activities that tempted him from his studies were the
theatrical productions that the Stift students presented at Shrovetide.
The subject was always either biblical or classical, and since there
were no women at the Stift, students like Johannes who were slight
of stature and not too loutish or clumsy played the female roles.
Johannes had the part of Mariamne in a tragedy about John the
Baptist. Unfortunately the play was performed in the open market-
place, and Shrovetide was in midwinter. That and the overexcite-
ment caused him to contract a "feverish illness"—one of many bouts
of bad health that threatened to hamper his studies.

Early in his university career, Kepler foresaw that theology and
mathematics, including astronomy, were always going to be linked in
his quest to discover for himself what was true and what was not. He
was fortunate to have Michael Mästlin for his teacher of mathematics and astronomy. Mästlin had won Tycho Brahe's admiration in
1578, when Tycho was collecting publications about the comet of
1577 through friends abroad. Mästlin's report stood out from the
others. Compared with Tycho's sophisticated methods, Mästlin's
were primitive: He had observed the comet by holding up a taut
string to line up reference stars and then looked up those stars in the
Copernican Prutenic Tables to find their positions. The results were
of great interest to Tycho, for both men had reached the same conclusion, with Mästlin's observations being slightly more accurate. In
a gracious letter written through a third party, Tycho suggested that
they exchange observations and indicated that Tycho would be
pleased to promote Mästlin's career in any way possible. At that junc-

Michael Mästlin, Kepler's mentor at the University of Tübingen and lifelong friend and correspondent.

ture, Tycho was willing to share his work and findings with other astronomers and engage in an exchange of ideas.

The University of Tübingen in Kepler's day still officially taught Ptolemaic astronomy, and Mästlin gave his students a good grounding in it. However, though he was a cautious man and far from outspoken on the subject, he was one of a mere handful of scholars in all Europe who believed that Copernicus's system of the cosmos should be taken literally, that the planets, including Earth, did in fact orbit the Sun.

Kepler also encountered the writings of Cardinal Nicholas of Cusa, who had insisted a century before Copernicus that Earth did not lie motionless in the center of the universe. Kepler reported, "I have by degrees—partly out of Mästlin's lectures, partly out of myself—collected all the mathematical advantages which Copernicus has over Ptolemy." Kepler soon came to agree with Mästlin, and he added a religious spin of his own to Copernican astronomy that made it seem to him even more likely to be correct.

In a universe created in the image of God, it made sense that the Sun, the brightest and most splendid of all objects, the source of light and warmth, should symbolize its Creator and be the center of all things. This was not an original idea. As early as the fifth century B.C. some Pythagoreans, in a pagan society, had thought similarly, except that they made the center of the universe not Earth or the Sun but an invisible fire. Classically educated Kepler was not ignorant of the Pythagoreans.

Kepler's idea went further. In both the Ptolemaic and the Copernican systems, the stars were in the outermost sphere. This sphere enclosed the universe and defined the extent of its space. To Kepler the sphere of the stars symbolized Christ, the Son of God. Kepler further reasoned that a sphere was generated by an infinite number of equal straight lines radiating from its center. Hence, the area between the symbol of God the Father, in the center, and the symbol of God the Son, encompassing the universe, represented the third member of the Christian Trinity, the Holy Spirit. In keeping with Christian doctrine of the Trinity, the three united in one. Neither in the spiritual universe nor in the physical universe could one of the three exist alone. Each required the others.

Kepler came openly to the defense of Copernican astronomy in two formal academic debates during his university years. He used his religious arguments along with others that sounded more like those of an astronomer. For instance, he argued that the time each planet takes to complete its orbit, and the planets' distances from the Sun, made better sense in the Copernican arrangement. The Sun was the source of all change and motion, and therefore it would not be surprising to find that the closer a planet was to the Sun, the faster it traveled—a line of thought that would prove fruitful for Kepler in the years to come.

From the Pythagoreans and Plato, as well as from neo-Platonic thinkers of Kepler's own era, he absorbed another idea that undergirded his preference for the Copernican system and guided his speculation and the course of his research from that time on. The philosophy that had impelled Copernicus to put the Sun in the center of the universe and had inspired the design of Tycho's Uraniborg was the worldview insisting that a profound hidden harmony, simplicity, and symmetry must surely underlie all the apparent complication and complexity of nature. This notion set fire to the spiritual and scientific imagination of young Kepler. A universe created by God could not be other than the perfect expression of such underlying order. What the goal of transforming astronomy was for Tycho Brahe, the search for this harmony in nature became for Kepler: an obsession that would occupy him for a lifetime.

Although during his student years Kepler worked busily and happily on astronomical questions and even wrote an essay about how the movements of the heavens would look from the Moon, he apparently remained oblivious to the possibility of pursuing any career other than theology. He passed the examination that signaled the end of his education in the arts, placing second among fourteen candidates, and received his master's degree in August 1591. He was nineteen. In a letter requesting that his scholarship be continued, the university senate paid him a tribute: "Young Kepler has such an extraordinary and splendid intellect that something special can be expected from him."

Kepler began the course of theological studies as something of a rebel, at least privately, for his earlier religious scruples continued. He now entered the realm of powerful men who opposed Calvinist teaching as ferociously as they did Catholicism. On such questions Kepler wisely chose to keep his thoughts to himself, not even sharing them with those mentors who most cherished him as a pupil. But in the privacy of his own mind, doubts about some Lutheran doctrines so oppressed him that he had to, as he put it, push aside all these complicated matters and sweep them completely out of his heart when he received Communion.

Meanwhile, the theological infighting that he now witnessed more closely as a student of theology in one of the leading schools so repelled Kepler that he grew to despise the entire controversy. He felt that such behavior was completely at odds with Christ's teaching, and he believed more strongly than ever that mutual tolerance between the divisions of the Reformation church was the only appropriate course. He was a devout Christian, and all his life this continued to mean that when it came to the intricacies of doctrine, he would not mindlessly accept the dicta of others.

However much Kepler's professors guessed about his views or shook their heads at his occasional attempts to defend Copernican theory, they nevertheless continued to recognize his promise. Kepler, though inwardly nagged by doctrinal doubts, had good reason to envision a smooth road lying before him and to imagine himself in clerical robes in a pulpit. Meanwhile, immersed in his studies, he was spending one of the happiest periods of his life. He would later conclude that whatever other forces came into play—exceptional insight into his talents, unfavorable judgment about his unorthodox views, or simple bureaucratic irrationality—it was definitely the will of God that brought an end to his happiness and set in motion a sudden and staggering change of plans.

The third theological year at Tübingen was probably a "holding year" of sorts, during which students who had completed their studies sought and found jobs. Just short of the end of this year, Kepler received the devastating news that his time at his beloved university was to end abruptly, and not in the way he had intended. A Protestant seminary school in Graz, Styria, in southern Austria, needed a mathematics teacher who knew history and Greek. The school appealed to the University of Tübingen, and Tübingen chose Kepler.

Graz seemed impossibly remote, in an area that was completely foreign to him. He had no plan or desire to be a mathematics teacher: He loved the subject and thought he might have a talent for it, but he considered himself not at all accomplished yet as a mathematician. He had been certain of his calling to be a pastor and serve his church. Surely, he thought, the move to Graz could not be God's will any more than it was his own.

On the other hand, Kepler's faith and his knowledge and experience of the way God guided the lives of men and women, as well as his sense of duty, told him that he must not be selfish. He hadn't been put into the world for himself alone. If it *was* God's will that he go to Graz and teach mathematics, he should not insist that he had a "higher calling." Furthermore, he had promised himself, when he had seen friends employing every device possible to avoid obedience when faced with similar distant postings, that he would be more dignified if it happened to him. Recalling that noble resolve, he admitted ruefully that he had thought he was "tougher than I actually was."

Kepler consulted his two grandfathers and his mother. Though disappointed that they would not see him in the pulpit, all thought it best to follow the advice of the Tübingen faculty. Kepler managed to engineer a compromise that left open the possibility of returning to service in the church, and with that settled he agreed to move to Graz.

The transfer was set in motion. Kepler could not go without the permission of the duke of Württemberg. Officials at Tübingen and the inspectors of the school in Graz sent letters requesting that Kepler be allowed to leave the duchy, and the duke gave his approval.

The twenty-two-year-old Kepler left his beloved Tübingen on March 13, 1594, with a heavy heart. The move to Graz was a venture into alien territory. Even the calendar was different: Because Württemberg still used the old Julian calendar, Kepler lost ten days when he came to the border of Bavaria, where they used the newer Gregorian calendar. He arrived in Graz, by that calendar, on April 11. Kepler trusted the will of God, but he could not have imagined how essential this strange, unexpected, seemingly senseless journey was in getting him to his future, to Tycho Brahe.

❦

WHILE KEPLER WAS BASKING in the rarefied scholarly atmosphere of the University of Tübingen, at the close of the 1580s and during the first years of the 1590s, the utopia of Uraniborg had begun to fray a little around the edges.

The Tycho who had built Uraniborg and Stjerneborg had usually enjoyed good relationships with most people. He did have serious conflicts with the peasants on Hven, but the customs of the time and the islanders' unusual immunity from those customs prior to his arrival made those conflicts almost inevitable. Otherwise, Tycho seems to have commanded genuine respect. He had founded and presided ably over an entirely unprecedented institution that drew both humble students and a scholarly elite from all over Europe and Scandinavia. Even in those days when an aristocrat could expect obedience, he had to have been a skilled manager. Students and assistants worked for him untiringly and apparently with great devotion. His ability to recognize the potential of men of lower classes and his willingness to elevate them to the level of valued colleagues—Steenwinkel being a case in point—set him apart from most of his aristocratic peers. He seems to have been a faithful husband to his commoner wife, concerned about his children, popular among commoner scholars and friends such as Pratensis, and well liked by royalty such as King Frederick and Wilhelm of Kassel. To a remarkable extent for a nobleman living at the end of the sixteenth century, Tycho had chosen to ignore the chasms between the social strata and had managed to bridge them.

His relationship with a new assistant who came to Hven in 1589 exemplified his continuing success in doing so. Longomontanus, as the man called himself, was the Latinized name for the farm where he was born in western Jutland to a poor peasant family. Poverty and the need to help his widowed mother run the farm had delayed Longomontanus's education, but the pastor of the local church had recognized his potential and seen to his schooling. Longomontanus was twenty-six by the time he entered the University of Copenhagen. Scarcely a year later, in 1589, he came to Uraniborg on the recommendation of his professors and was soon one of the most skilled and exacting astronomers there. His humble origins did not prevent his becoming one of Tycho's favorites and an intimate friend of the family. Tycho trusted him sufficiently to make him his personal secretary. Longomontanus was popular with Tycho's two sons, Tycho and Georg, and Tycho chose him (perhaps because of his maturity) to chaperon them when they traveled to visit relatives.

Nevertheless, in spite of his lack of regard for social divisions and the scorn he had poured on the nobility, Tycho had not ceased to be an aristocrat. It was in his blood and upbringing. If for him the traditional nobility had stopped being much more than a weary charade, the symbolism of Stjerneborg indicated that it had been supplanted by a much more vigorous and much older aristocracy, in which Tycho saw himself the heir by divine right to Hipparchus, Ptolemy, and Copernicus, men who left kings like Frederick in the dust. Nor had Tycho shed the trappings and pride of nobility. He had transformed them, with embellishments, to adorn this different kingdom. Here, the old social class distinctions really had, to a certain extent, become extinct, but Tycho was still on the top—higher, in fact, than he had been in the old order, and potentially more of a tyrant.

The picture of Tycho that has chiefly come down in history is not a sympathetic one, and a few years later he would appear mercurial, autocratic, and paranoid to Kepler, who was initially inclined to revere him. The finer side of Tycho never disappeared completely, and previous instances of tyranny and paranoia may have escaped the records, but evidence points to a change in his temperament in the early 1590s. In the autumn of 1590 he imprisoned his tailor for three days before the man succeeded in escaping from Hven by night. In 1591 Tycho's jester, a dwarf named Jepp, tried to flee from Uraniborg, and Tycho had him beaten. None of that was perhaps out of character for the lord of a fief, but as the decade continued, there was increasing evidence of a less likable aspect to Tycho's personality.

In the spring of 1591 Tycho's concentration on astronomy suffered serious disruption from an embroilment with a gentleman tenant, Rasmus Pedersen, who lived on a small manor in Zealand that was part of Tycho's holdings as a canon of Roskilde Cathedral. The incident began when Pedersen purchased from Tycho a life tenancy of Gundsøgaard, an unprofitable estate with only nine cottages and the ruins of a manor house that had recently burned down. Tycho cannot have charged him very much for this holding, and he was chagrined when Pedersen unexpectedly turned it into a thriving establishment and rebuilt the house into something large and quite splendid. Pedersen may or may not, in the process, have overstepped his fishing rights and exploited his nine peasant families beyond the usual norm. There were reports that after using their free labor for his building project, he forcibly ejected them from their small plots so that the land could become fields of the manor, then used their free labor again to build new cottages, and finally obliged them to buy the cottages from him.

Motivated either by concern for the peasants or a desire for the manor house, or both, Tycho attempted to renegotiate the lease and, when Pedersen refused, seized the manor and expelled him without a refund. But Pedersen was persistent: When Tycho ordered the fields at Gundsøgaard to be plowed and sown, Pedersen had his retainers sow fifty-two and a half bushels of rye right behind Tycho's sowers. Tycho escalated the conflict. His men seized Pedersen as he was dining and brought him in irons to Hven, while others of Tycho's agents confiscated Pedersen's business records and detailed reports about the estate. Tycho imprisoned Pedersen at Uraniborg for six weeks until Pedersen agreed to sign a capitulation. By the spring of 1591, Tycho had brought the matter to court.

A few months later, Tycho's appeal to the king and the Rigsraad failed when even Tycho's aristocratic peers, who might have been expected to side with one of their own, were unwilling to view his treatment of Pedersen as normal behavior for a feudal lord. On the way home to Hven, the disgruntled Tycho voiced his disappointment by composing a Latin epigram complaining of this unfair decision. He did not drop the matter. His next maneuver was to try to link Pedersen with a drowning in a well. That failed. By November 1592 Tycho was holding Pedersen's brother and a servant as prisoners. There is no record of the result of a second hearing, for which Tycho was allowed to nominate some panelists to participate in the decision, but two years later he seems to have regained possession of the manor house of Gundsøgaard. There are no further records concerning Pedersen.

Possibly while Tycho was holding Rasmus Pedersen prisoner at Hven, a young man named Georg Ludwig Frobenius arrived. As Frobenius described events in his memoirs,* he had received his master's degree from the University of Wittenberg, worked for a year as a tutor in Saxony, and gone to Denmark, afire to visit Uraniborg and meet Tycho Brahe. He made the initial mistake of seeking entrance after everyone had gone to bed. He might have been forgiven for thinking that since this was an astronomical observatory and it was "a beautiful, clear, and calm night," the entire household would not be asleep. However, despite the glowing letters of introduction that Frobenius presented, the porter at the gate turned him away. With the savage barking of the mastiffs kenneled above the gatehouse ringing in his ears, Frobenius walked off to spend the night hungry in a field.

*Frobenius's story remained unknown until John Robert Christianson discovered his memoirs in the late 1980s.

Early the next morning Frobenius tried again at the gate, and Tycho granted him an audience. After conversing with Frobenius and reading his letters of recommendation—one of them from Tycho's friend Caspar Peucer—Tycho agreed to accept Frobenius as a student. He would have free bed and board and an assigned seat at the dinner table.

Though several languages were spoken at Uraniborg, Danish was the most common, and Frobenius knew no Danish. From the start he felt excluded, particularly at meals, where he was seated beside a student from Bergen, Norway. However, things must have gone rather well, for about a month after his arrival Tycho asked him, through other students, whether he would like to remain at Uraniborg "to serve him in the study of astronomy." Frobenius replied enthusiastically. He would be pleased to stay for one, two, or three years, if the terms of employment were acceptable. Since most contracts for service at Uraniborg were for three years or less, Frobenius had not asked for special favors.

Nevertheless, at this point there was a mysterious alteration in Tycho's attitude toward Frobenius. The conditions of the contract, as Tycho laid them down, seemed intentionally framed to make it impossible for Frobenius to accept. Frobenius was shocked and somewhat affronted to hear that one, two, or three years were out of the question. He would have to commit himself for six years minimum, pledge never to reveal anything about Tycho's inventions to anyone either now or after he left, take no notes for his own benefit or later personal use, and "serve without hesitation wherever [he] could fruitfully be used, in any of [Tycho's] astronomical or pyronomical labors." He would eat at Tycho's table, but there would be no salary or clothing provided by Tycho, who preferred "to grant to me whatever happened to come my way."

Frobenius took a deep breath and asked for time to consider. Then he replied that he could not agree to such a long period as six years. He wanted to visit foreign lands and learn foreign languages—

as Tycho himself had done—and eventually probably pursue a career in medicine or law. Though he was willing to promise not to reveal or spread abroad any information about Tycho's inventions or observations, he was reluctant to promise never to utilize knowledge gained at Uraniborg to benefit his own studies, for it would be a waste of time to learn things he could never use in the future. Also, he needed a fixed salary, not just bed and board.

Contract negotiations with prospective assistants were not unusual, and Tycho was often willing to bend on the length of service and adjust other clauses. Not so in the case of Frobenius. Meanwhile Frobenius learned that his situation was disturbingly paradoxical, for though Tycho's terms of employment seemed designed to force him to leave, other students and assistants told him that it was extremely difficult to get away from Uraniborg. Hans Crol, who was, like Frobenius, German, said it would be impossible to escape the island unless he could find a good pretext.

The distraught Frobenius recalled that he had in his possession letters of reference to other people besides Tycho, one of which might provide him with an excuse to take a leave of absence from which he would simply not return. The letter that looked most promising was to Heinrich Rantzau in Holstein. Frobenius requested leave for only a few weeks to go to Holstein, claiming to have been entrusted with an oral message to Rantzau that could not be delivered by anyone else. At first Tycho, perhaps suspecting Frobenius's intentions, refused his permission. Frobenius then offered to leave all his belongings in his trunk at Uraniborg as security for his return. Tycho finally acquiesced, with the strange requirement that Frobenius seal the trunk on all sides. Tycho and Frobenius traveled together to Copenhagen. When they parted so that Tycho could attend a meeting of the Rigsraad, Frobenius found a ship bound for Lübeck, hurried aboard, and sailed away with only "a couple of shirts, a cloak, and handkerchiefs in a black linen satchel."

It is tempting to wonder whether some of Tycho's assistants were playing a practical joke on poor Frobenius, whether he made the story up, or whether he was perhaps a difficult and vindictive person himself. Tycho in fact allowed many students and assistants to leave Uraniborg for posts elsewhere. Most continued to be his good friends, and he viewed it as an advantage to have a network of them all over Europe. Crol, who warned Frobenius about the difficulty of escaping, was at the time an embittered man because of the recent death of his son. Crol himself never left Uraniborg and died in the autumn of that same year. Tycho grieved for him and praised his memory as a fine goldsmith, instrument maker, and observer.

Conflicting opinions and reports are, of course, not unusual about men and women whose lives for one reason or another tower over the people around them. On the one hand there are those who revere them and either do not experience or choose to forgive treatment that others regard as insulting or abusive. On the other hand there are those who, wearing different spectacles or having somehow inadvertently fallen foul, experience that greatness as having a nasty side indeed.

11

YEARS OF DISCONTENT

1588–1596

IN THAT SAME SPRING of 1591, when Pedersen was in Tycho's dungeon and Frobenius was upstairs plotting his escape from Uraniborg, Tycho received a letter from Wilhelm, the landgrave in Kassel, asking about an animal that Wilhelm called a Rix. Wilhelm had heard that a Rix was taller than a deer and native to Norway, and he inquired whether Tycho might have a picture painted of a Rix and sent to him. Tycho suspected that the landgrave meant a reindeer and was hinting that Tycho might send not merely a picture but the animal itself. Tycho did not have a reindeer, but he had an elk, so he offered to send that. No, the landgrave replied, he already had an elk, and it was not a reindeer he wanted either. He had had one of those before, and it had not survived in the climate of Kassel. A Rix was what he wanted. Nevertheless, he would not turn down an elk or two if offered.

Tycho had an elk brought from Norway to Copenhagen, where it was to wait at his niece's home until it could be shipped. Unfortunately, the elk mounted the steps of the manor house, got into the beer supply, and consumed so much beer that it fractured a leg trying to get back down the stairs and died.

In the course of correspondence with Wilhelm around this pitiful story, the first hint came from Tycho himself that he was not entirely happy. He went so far as to tell Wilhelm in rather cryptic language that he might choose to venture into other climes, the sky above being available for study anywhere. He did not specify what was troubling him. Perhaps it was only a temporary low mood or annoyances like Pedersen and Frobenius that he had largely brought on himself. But there is reason to suspect that one cause of Tycho's discontent, and his poor handling of those annoyances, was something more significant—a crushing disappointment in his astronomy.

Tycho did not, in this correspondence with Wilhelm, learn of Bär's visit to Kassel, but Bär had never been far from Tycho's mind. Tycho's book about the comet, with the chapter about the Tychonic system, had come out in April 1588. He mentioned nothing about the success of the 1587 parallax observations in the book, but not long after its publication he wrote to his friends Caspar Peucer and Christoph Rothmann, saying that he had observed the parallax *all the way back in 1582*. That claim flatly contradicted an earlier letter that he had written in 1584 to a professor with whom he had studied at Rostock, Heinrich Brucaeus. Tycho had reported to Brucaeus that in the observations made in 1582, he had been unable to find a parallax for Mars, and that the Copernican hypothesis had therefore to be rejected. In 1587 he had seemed a little less certain of that negative result when he had written Wilhelm, in Kassel, that (based on those same 1582 observations) he was more confident he would find a parallax. Then, in 1588, he told Peucer and Rothmann (still referring to the same 1582 observations) that he had finally succeeded. In November 1589 he repeated that claim in a letter to Thaddeus Hagecius, stating once again that all the way back in 1582 (two years before his 1584 letter to Brucaeus saying the opposite) he had observed a diurnal parallax and that it had been large enough to convince him that Mars comes closer to Earth than the Sun does. In the same letter to Hagecius, Tycho also spoke of the Tychonic system that he had "thought out . . . very nearly six years ago."

The puzzle was, and is, why the 1582 observations were so important to Tycho that he kept harking back to them, making contradictory claims about their results. He had designed and built several instruments after 1582 that were much more capable of making this measurement. Kepler later examined Tycho's observations of 1582 and reported that he could find no evidence in them for a Mars parallax. Subsequent knowledge of the solar system has confirmed that there could have been no evidence of one there. Yet in 1588 and 1589 Tycho's letters insisted there was.

The explanation for the contradictory claims* almost certainly had to do with Bär. Six months after Tycho published the book with the chapter about the Tychonic system in April 1588, Bär also published a book entitled *Fundaments of Astronomy*. In it he laid out a system he claimed was his own invention. Except for some details, it was identical to Tycho's system. Tycho's nightmare had come true. It was of utmost importance to establish priority, and to make it indisputably clear that Bär had learned of the system from Tycho, not Tycho from Bär. One way to accomplish this was to prove that not only the idea of the system, but observations to demonstrate its superiority, predated Bär's visit to Hven. Certainly this provided ample motivation for letters to other scholars in which Tycho made claims about the 1582 observations. In fact, the date itself was far more important than the findings, and that was fortunate.

In the autumn of 1589, after he had written to Hagecius, Tycho almost immediately turned his attention again to the problem of refraction. He did so with considerable trepidation. He had complete confidence in his Mars observations of 1587, but he knew he did not have as much evidence to support the use of his table of solar refraction, the table he had used in 1587 when correcting the positions of Mars to take refraction into account.

---

*Gingerich and Voelkel point out the mysterious contradictory sequence of letters and explain how they have arrived at this explanation in "Tycho Brahe's Copernican Campaign."

The result of Tycho's new study of refraction was heartbreaking. He found that his solar refraction table was badly flawed. He had, in fact, found *no* parallax for Mars in 1587. Tycho never again claimed he had discovered a parallax for Mars, nor did he make any more serious attempts to find it. The oppositions of Mars in 1591 and 1593 took place in summer, when the nights were too short to try to measure a parallax.

❧

WHILE TYCHO UNDOUBTEDLY MADE some of his problems worse by mishandling, in other cases he acted with extreme patience and care for his family. The sad love story of his sister Sophie was one of the latter.

Twelve years younger than Tycho, Sophie had always been a favorite. When she was a child, he had taught her some astronomy, and, when she was fourteen, she had assisted him in the observation of a lunar eclipse at Herrevad. Since then Sophie had married a rich nobleman, borne a son, and lived in splendor at his Eriksholm Castle.

In 1588, Sophie's husband died, and with her own family inheritance as well as her husband's she was left a wealthy young widow. Sophie was a frequent visitor to Uraniborg in the autumn and winter of 1589, and it was there that she met Erik Lange, the well-educated, well-traveled young gallant who had first brought Bär to Hven as part of his retinue in the autumn of 1584.

Sophie fell in love with Lange, an exceedingly foolish choice for a wealthy widow, for Lange's own considerable fortune was rapidly vanishing, squandered to support his obsession with a futile branch of alchemy devoted to trying to turn base metal into gold. In 1590, the same year he and Sophie Brahe were betrothed, over the objections of all Sophie's siblings except Tycho, Lange had been forced to sign over his estate of Engelsholm to make good his debts, and even that had not been enough to cover them. He was placed under house arrest. Though creditors could not touch the substantial

Tycho's sister Sophie Brahe, in an oil painting by an unknown artist.

part of Sophie's deceased husband's estate that was held in wardship for their son, soon after the betrothal they began closing in on her personal fortune.

In 1592 Sophie's and Lange's problems became, more than ever, Tycho's when Lange fled in secret to Hven, and Tycho helped him escape from Denmark and his creditors. Lange's addiction to the dream of turning base metal into gold was beyond control, and he continued to run up enormous debts, dodging from place to place in Europe. Sophie, distraught and besieged by his creditors at her home at Eriksholm Castle, often sought refuge at Uraniborg. In spite of having followed her heart where no sensible person should have allowed it to lead, Sophie was a strong, quick-witted, self-confident woman, involved in numerous intellectual pursuits and highly respected by many people, not least Tycho. He considered her one of the most intelligent and learned women he knew and he enjoyed her company. Nevertheless, her problems and Lange's continuing disastrous foolishness were a drain on his patience and energies.

WHILE TYCHO WAS RISKING loss of respect from his peers over the Pedersen affair and breaking the law to help his brother-in-law, he was treading even more perilously in another respect. The estate attached to the Chapel of the Magi at Roskilde was extensive and provided Tycho with a large annual income. Nevertheless, he had neglected his responsibilities and allowed the chapel to fall into disrepair. Since 1591 he had received repeated requests from the boy-king Christian to repair the leaking roof. Tycho ignored him. There seemed little to fear. Christian had not yet reached majority, and his limited powers allowed him to do little more than play at being king. Furthermore, the Regency Council that ruled during his minority was packed with Tycho's friends and relatives. Christian's father Frederick had been so enthusiastic and supportive of Tycho that Tycho had forgotten that kings needed extremely sensitive treatment and could not safely be regarded as familiar equals or, if they were still minors, as irritating nephews.

Tycho had every reason to expect Christian to be as supportive as his father had been. Christian made a royal visit to Hven in July 1592. Not only was the weather beautiful and the banquet, wines, music, conversation, and the humor of Tycho's jester all very much to the king's liking, but he was fascinated with Tycho's instruments and the many treasures of Uraniborg.

Though Christian's memories of Hven would in fact remain fresh with him all his life, in the summer of 1593, a year after his day on the island, Christian visited Roskilde Cathedral, and his attitude toward Tycho Brahe changed catastrophically. Christian found the roof of the Chapel of the Magi in such a state of near collapse as to threaten the alabaster and marble sepulchers of his father and grandfather. The young monarch dictated an angry message demanding that Tycho begin repairs. If he failed to do so, Christian himself would hire a builder and have the work done at Tycho's expense. The message commanded a reply by return courier.

Tycho must indeed have been distracted by other matters to have ignored this ultimatum. By the following autumn, 1594, there had still been no repairs, and Tycho received another angry complaint ordering him to have the work completed by Christmas or forfeit the fief with its incomes. This message succeeded in getting Tycho's attention, but, still seemingly unaware of the dangerous path he was treading in his cavalier treatment of Christian, Tycho had the roof rebuilt as a flat one with wooden beams rather than restore the original vaulting.

The problem that seemed much more urgent and troubling to Tycho in the autumn of 1594 was the arrangement of a marriage for his eldest daughter. Plans had begun happily the year before. Tycho's four daughters—Magdalene, nineteen, Sophie, fourteen, Elisabeth, thirteen, and Cecilie, twelve—were wealthy, accomplished young women. Because their mother was a commoner, they could not marry into the nobility, but Tycho had every reason to believe that he could arrange matches so that their standard of living would be equal to what it had been at Uraniborg. Among the academic elite there were young professors and physicians of high social status who were as well versed in late Renaissance culture as noblemen, often more so. Clearly, Tycho Brahe's daughters were enormously good catches for any of these.

In May and June of 1593, a young man named Gellius Sascerides became a prime candidate for Magdalene's hand. Gellius came from a prominent academic family. His father, though Dutch, was a distinguished professor in Copenhagen, and the family and Gellius himself had a wide circle of important friends there. Gellius and his brother David had both been assistants at Uraniborg, Gellius for five years when he was in his early twenties. He had been one of Tycho's most promising young disciples. Tycho's daughters had been children then. Magdalene was only about eight when he arrived and thirteen when he departed in the midsummer of 1588, but they must have met at the dinner table. Gellius had continued to serve as Tycho's representative to foreign courts and universities. Now he was back, thirty-one years old—well traveled, brimming with confidence, with a medical degree and a reputation as a rising star among the young scholars of Europe. There could not have been a more appropriate suitor for Tycho Brahe's eldest daughter.

According to the courtship and marriage customs of late-sixteenth-century Denmark,* the procedure did not begin with the couple themselves but with tentative, informal feelers among friends, relatives, and the broader network these could provide. If a situation looked encouraging, a young man asked trusted friends to act on his behalf and present his proposal to the prospective bride's immediate family. Gellius chose his friend Mogens Bertelsen Dallin as his emissary. After preliminary discussions with Tycho, Dallin and Gellius went to Hven with a formal proposal on September 19.

Custom required the suitor's spokesman to deliver a long, elegant speech, climaxing in the formal proposal of marriage. The suitor and his party then took seats, and the bride's family replied. Tycho's sister Sophie probably was the spokesperson for the Brahes when this ceremony took place at Uraniborg, for it was she who went on to negotiate on their behalf. Her speech would not have answered yes or no: At this point the suitor was committed, but the bride's family was not.

After that, the negotiations began, and almost immediately there were difficulties. Although Magdalene was not of the nobility, she had been raised as a noblewoman, and her immediate family was far wealthier than Gellius's. Gellius had no substantial fortune and no

---

*The description of these customs comes from social historian Troels Frederik Troels-Lund, as redescribed in English by John Christianson.

permanent position providing him an income, only his prospects as a brilliant young scholar. There was a huge discrepancy between Tycho's and Gellius's expectations of what the wedding and the couple's living style would be like.

Magdalene's family was of course well able to supply a suitable dowry, including not only money but such necessities as enough fine gowns to last a lifetime, jewelry and embroidered caps, table and bed linens, kettles and cooking pots and pans, tapestries, a bed and all its hangings, duvets, bolsters, and pillows. All of this was laid out in the marriage contract. By custom a young woman entered marriage supplied for life, leaving her husband responsible only for day-to-day needs.

Nevertheless, Gellius must soon have realized that he was out of his depth. Even if he had anticipated Tycho's plans that the marriage take place on the scale of a noble alliance, he had no experience of how costly that would be. The wedding itself, although most of the cost fell to the bride's family, would be staggeringly expensive for him. A nobleman's wedding lasted five to nine days, and the groom needed new, sumptuous clothing every day. Gellius's entire inheritance, and then some, could easily have been eaten up in supplying a betrothal gift—customarily a number of small silver items and a massive gold chain that ensured his wife could support herself for life merely by pawning it—and a "morning gift," which might be anything from valuable jewelry to the hereditary rights to landed estates. Clearly it was the duty of Dallin, Gellius's spokesman, to scale down these expectations as he negotiated the contract.

When it came to a way for Gellius to support his bride, Tycho proposed that Gellius reenter his service at Uraniborg, no longer as an assistant but as a son-in-law, subordinate only to Tycho himself. This arrangement had been part of the nuptial contract agreed on at Uraniborg in late September. Tycho may well have been considering Gellius as a prime candidate to succeed him either as director of Uraniborg or as codirector with one of Tycho's sons.

Thus matters stood when Tycho received his ultimatum from Christian about the Chapel of the Magi and went to Roskilde, taking Steenwinkel and another builder with him. He was away again for a short period in late October, and when he returned to the island there was a letter waiting for him from Sophie Brahe in Copenhagen. Gellius had requested an amendment to the contract so as to have only a small wedding with a banquet for a limited number of guests, not a celebration lasting for days, and he would agree to remain in Tycho's service only until the following Easter, no longer—which seemed astoundingly unappreciative of what had struck Tycho as an extraordinary offer to the young man. It is testimony to Tycho's concern for Magdalene that he swallowed his annoyance and sent word to Sophie that he would accept Gellius's amendments but hoped he would not go on changing his mind.

However, for reasons that still remain mysterious, Gellius had grown extremely skittish about an alliance with Tycho Brahe's family. He continued to raise fresh objections and make new demands. The renegotiations continued, then floundered and fell apart entirely in late November, with Gellius frustrating every effort at reconciliation. Ominously, his remarks about the Brahes in public and among his close friends seemed contrived to bring an irreversible end to the marriage arrangements.

On December 12 Tycho formally declared the wedding contract canceled. A devastated Magdalene wrote a statement describing her mistreatment in this matter by Gellius. The breakdown at this point in marriage negotiations, when the man and woman as a formally betrothed couple could by custom already have slept together, was like a death knell for her. No man would ever marry her. At the age of twenty, having had every prospect of happiness and family before her, Magdalene knew that her erstwhile friend and suitor Gellius had condemned her to spinsterdom.

The matter had by now become more than a private concern and threatened the Brahe family with public disgrace. Professors at the

Tycho Brahe in the portrait from his book *Astronomiae Instauratae Mechanica*, 1598, printed shortly before he went to Prague.

Johannes Kepler in a portrait painted in 1610 when he was at the height of his career as Imperial Mathematician in Prague.

Knutstorps Borg, Tycho Brahe's birthplace and ancestral manor house, as it appears today and (*below*) in a seventeenth-century drawing by Gerhard von Burman.

Left: King Frederick II of Denmark, Tycho's patron, who granted him the fiefdom of Hven and whose support enabled him to build Uraniborg, in a portrait by an unknown artist.

Opposite above: Tycho's elevation drawing for Uraniborg.

Opposite below: Tycho's plan of the gardens of Uraniborg, with the mansion at the center.

ORTHOGRAPHIA PRÆCIPVÆ DOMVS ARCIS VRANIBVRGI
in Infula Porthmi Danici Venufia, *Vulgo* Huenna, Aftronomiæ inftaurandæ gratia, circa annum MDLXXX,
à TYCHONE BRAHE, exædificatæ.

ARCIS VRANIBVRGI, A TYCHONE BRAHE, DÑO DE KNVDSTRVP,
IN INSVLA HELLESPONTI DANICI HVENNA CONSTRVCTÆ, QVO AD TOTAM CAPACITATEM, DESIGNATIO.

Above: The great globe, begun in Augsburg in 1570 and completed at Uraniborg a decade later. It became the centerpiece of Uraniborg's library.

Right: The great mural quadrant, one of Tycho's most splendid instruments, built into the structure of Uraniborg in 1582. Tycho considered this portrait of himself an excellent likeness. The two men shown in the foreground were not part of the mural; they were part of this particular drawing of the mural.

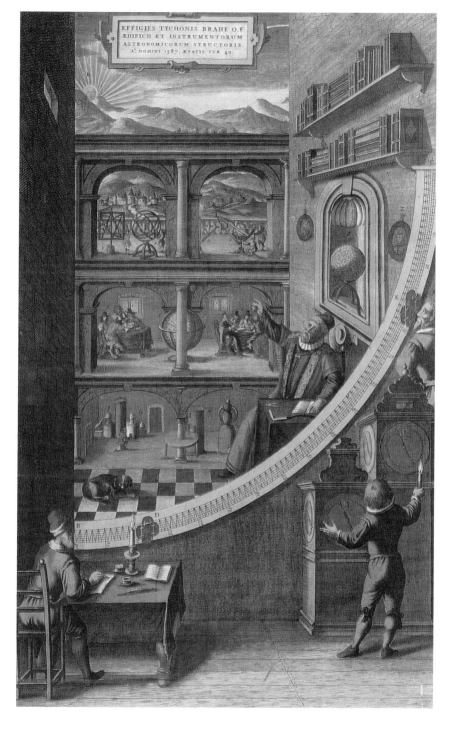

EFFIGIES TYCHONIS BRAHE O.F.
ÆDIFICII ET INSTRUMENTORUM
ASTRONOMICORUM STRUCTORIS.
A° DOMINI 1587, ÆTATIS SUÆ 40.

Stjerneborg ("Star Castle"), the partly subterranean observatory Tycho built in the 1580s outside the perimeter wall of Uraniborg to accommodate some of his largest and most powerful instruments.

The Chapel of the Magi at Roskilde Cathedral, with tombs of Frederick II and his father, as it appears today. It was this chapel that Tycho neglected to keep in good repair, thus incurring the wrath of the teenage Christian IV.

King Christian IV of Denmark, whose birth horoscope Tycho drew up, and who later became Tycho's nemesis, in a portrait by Pieter Isaacsz.

Wedding medallion portraits (1597) of Johannes Kepler and his first wife, Barbara.

Above: The cliff-top Benatky Castle, northeast of Prague, that Emperor Rudolph gave to Tycho to create a second Uraniborg. The mural depicting hunting scenes and featuring the emperor is visible on its walls.

Above left: Emperor Rudolph II of the Holy Roman Empire, the eccentric, reclusive patron of both Tycho and Kepler, in a portrait by Hans von Aachen.

Above: Drawing by an unknown artist depicting riots between Archduke Leopold's troops and Protestant vigilantes in the streets of Prague, near the Keplers' home, in February 1611.

university mediated a new marriage contract in January in which Gellius would not reenter the service of Tycho at all. To enable him to afford the obligations of married life, he was to be appointed provincial physician of Skåne, on petition of the nobility of that province, many of whom were Tycho's kinsmen. Gellius reneged once again, and this time he judged it prudent to shift the blame by openly spreading malicious rumors about Magdalene and her family and attempting to sow dissension among Tycho's relatives. The final breakdown of all negotiations occurred before the next autumn, 1595.

The gossip surrounding this affair was catastrophic for Tycho. He watched helplessly as the high esteem in which he had thought Danish society held him evaporated like a mirage and he became a laughingstock. Gellius had found that preemptive slander was a very effective way to defend his own reputation, and he used that weapon on every possible occasion. From the court in Copenhagen and all parts of the kingdom, Tycho heard reports of people sniggering at him, exchanging ribald jokes about the daughter he loved, sneering at Kirsten's common origins, blaming Tycho and Sophie for failing to negotiate in good faith, dragging all their names through the mud. Ill will in the university, kept under wraps while Tycho appeared impregnable, emerged to add new voices to the din of calumny.

By October 1595, when Mars was once again in opposition at a more favorable time of year to attempt the parallax observation, Tycho barely noticed, leaving the work to Longomontanus and others on his staff. They obtained both morning and evening observations only once, on October 27, and none of the observations are written down in Tycho's hand.

In January 1596 Tycho took the only step that could restore his family's reputation. Though he had decided repeatedly not to do so for fear of adding to their grief, he at last brought suit against Gellius for breach of contract. Hearings took place in Lund and then in Copenhagen. From there the case was referred to the diet of the Danish nobility and then reassigned by the crown a year later to a special court of nobles. The specific outcome of the trial is unknown, except that Gellius seems not to have suffered by it. He was soon granted a living of one canonry and two vicariates in Lund Cathedral and later became a professor of medicine at the University of Copenhagen.

Tycho was fearful that the loss of esteem caused by the failed betrothal would trigger similar humiliation for him in scholarly circles even beyond the borders of Denmark. He decided it would be a wise move to enhance his reputation as an astronomer and deter his rivals by publishing an anthology of his correspondence having to do with astronomy. *Epistolae Astronomicae* (Astronomic Letters), which he printed in his new paper mill, clearly revealed his belief that Bär had plagiarized his planetary system. He included the letter he had written to Rothmann about the success of the observations of 1582. Whether or not those observations showed a parallax for Mars, they proved that he had worked on the problem of Mars and on developing his planetary system long before Bär came to Hven.

In the dedication carved on the cornerstone of the paper mill, Tycho's words indicate how stubbornly independent he was feeling:

> This dam and paper-mill with all their accessories and the fish ponds above them have been built on the order of Tycho Brahe on a site where nothing of the kind existed previously, by his own design, under his own supervision, and at his own expense for the benefit of his country, himself and his heirs. Let us do good while we have the time.

Even while he thus reminded Denmark that his work had been for his country as well as for himself, Tycho was becoming increasingly resigned to leaving. Early in May 1596 he ordered that his pilot-boat be refitted so that it could serve as a cargo boat. Clearly, he was thinking of transporting unusually heavy items either to or from Hven.

1 2

GEOMETRY'S UNIVERSE

1594‑1597

IN GRAZ, where Kepler reluctantly began his teaching job in 1594, he found a much less stable religious climate than in solidly Lutheran Tübingen. Here and in the rest of Styria, Protestants and Catholics lived side by side in nominal peace, but only thinly disguised their hostility. The rulers of Styria were members of the royal Hapsburg family and staunch Catholics. Under the Peace of Augsburg they had the right to declare that everyone in Styria must be Catholic. However, nearly all the most powerful landholders were Lutheran, and the Hapsburgs had found it advisable to allow Protestant nobles in the countryside and Protestant citizens in Graz and other cities to practice their faith openly.

The school where Kepler taught was thick in the middle of this uneasy truce and by no means neutral. It had been established in 1574 as a deliberate countermove to the founding of a Catholic Jesuit college the previous year. Its four preachers and twelve teachers were influential members of the Protestant community, and the school was, to all intents and purposes, the rallying center for Protestants in Graz.

At first none of this controversy touched Kepler. His concerns

Graz, in an early seventeenth-century engraving attributed to Matthäus Merian Sr.

were almost entirely academic. His teaching duties were in the upper school, where he taught advanced mathematics, including astronomy. Either he was not a particularly exciting teacher or the subjects were not popular, because few students attended his classes in the first year and none in the second. School officials looked for other classes Kepler could teach and assigned him rhetoric, Virgil, less advanced arithmetic, ethics, and history.

Kepler's duties extended beyond the classroom. He was also district mathematician, a public office that had considerable responsibilities connected with it. One of them was the compilation of an annual calendar with astrological predictions for the coming year. Kepler could approach this task with confidence. Already at Tübingen he had been known for his astrological skills. The district mathematician's calendar told what to expect about weather, harvests, war, disease, the most auspicious periods during which a physician might bleed a patient or perform surgery, when farmers should sow seed, when the weather would be most benign or most inclement or dangerous, when the Turks would attack, and when one should anticipate religious or political troubles. What Tycho had to produce for the princes of Denmark, Kepler had to produce for the entire citizenry of Graz and the surrounding countryside—albeit for only a year at a time instead of a lifetime.

Kepler's attitude toward astrology fell far short of the confidence most of his contemporaries had in it. He was already calling it the "foolish little daughter" of respectable astronomy. Later he would write that he abhorred "nourishing the superstition of fatheads" and that "if astrologers do sometimes tell the truth, it ought to be attributed to luck." However, he did not fully reject the idea that there were links between the cosmos and human beings. He, like Tycho, thought that the movements of the planets must in some way influence what happened on Earth, but probably far more subtly and far less deterministically than was commonly supposed.

Whatever his reservations, Kepler produced the calendars; they were, after all, a part of his job description. Before long, Graz found it had a very able district mathematician indeed, though he was not exactly a bearer of good news. For 1595, he had predicted an exceptionally cold winter, an attack by the Turks from the south, and a peasant uprising. All of those prophecies came true.

Michael Mästlin, Kepler's mentor at Tübingen, was particularly scornful of Kepler's astrological activities. Kepler took exception with him in a letter: "If God gave each animal tools for sustaining life, what harm is there if for the same purpose he joined astrology and astronomy?"

However, Kepler, the astrologer, failed to predict the most momentous event in his own life that year, a discovery he made in his classroom. Until then, Kepler had been an obscure teacher with some mathematical skills and little to set him apart from hundreds like him in Europe. Now emerged the Kepler who would transform astronomy, and also the Kepler whom some would think quite mad, for the discovery he made that day sounds to the twenty-first century almost as outlandish as the astrology for which he was admired.

On July 19, 1595—he kept careful record of the date, so significant did it seem to him—Kepler drew a diagram for his students on the chalkboard. The drawing demonstrated the progression of the Great Conjunctions of the planets Jupiter and Saturn; that is, when Jupiter passes Saturn in the zodiac. Jupiter and Saturn are the slowest moving of the planets that were known in Kepler's time, and since Tycho Brahe had recorded the Great Conjunction of 1563 as the second entry in his logbook, there had been only one other.

To understand what Kepler drew, picture the celestial sphere with Earth at the center. The two planets, Jupiter and Saturn, travel in enormous circles around Earth. Every twenty years Jupiter catches up with Saturn, the more distant of the two planets, and passes it. Kepler's drawing showed that these passings, or conjunctions, do not happen every time at the same points in the zodiac. For example, the fourth conjunction in the drawing (1643) occurred at *almost* the same point as the first (1583), but not quite; the fifth at almost the same point as the second, but again not quite, and so on. As Kepler drew the lines, each went just beyond joining a former line to make a closed triangle. Instead, the quasi-triangle "rotated," creating the pattern on the chalkboard. In Kepler's words, "I inscribed within a circle many triangles, or quasi-triangles, such that the end of one was the beginning of the next. In this manner a smaller circle was outlined by the points where the lines of the triangles crossed each other."

That second circle was visible in his drawing, half the size of the outer circle. While the *points* of the triangle "drew" the outer circle as the triangle rotated, the *middles of the sides* of the triangle "drew" the inner circle. The triangle's lines never trespassed into the inner circle, and its points never went outside the outer circle. The triangle dictated how far apart the two circles had to be.

Looking at what he had drawn, Kepler was struck by an insight that made him feel as though he had suddenly and unexpectedly opened a book and found inscribed there the secret of creation. As he wrote later: "The delight that I took in my discovery, I shall never be able to describe in words."

It was in Kepler's nature that as soon as he had resigned himself a

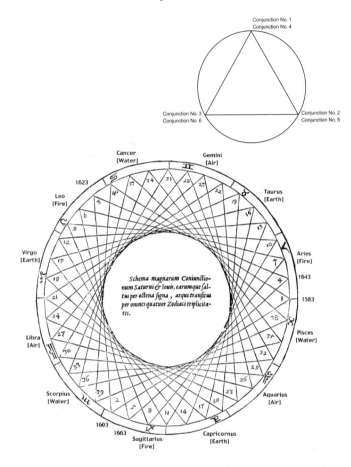

Figure 12.1: The pattern of Jupiter-Saturn conjunctions, showing where they happened in the zodiac. The conjunction in 1583 (right side of drawing) occurred when the two planets were in Aries/Pisces. The conjunction in 1603 (lower left) was in Sagittarius, in 1623 in Leo, in 1643 in Aries, in 1663 in Sagittarius, and so on. The drawing is from Kepler's *Mysterium,* with the zodiac names and the dates added around the rim. The reason for the names of the elements—earth, air, fire, water—will be explained later in the discussion of the appearance of "Kepler's Star."

If the conjunctions occurred repeatedly in the same positions in the zodiac, Kepler's drawing would have looked like the insert, upper right. Instead, they "progress," as represented in the central figure.

year earlier to teaching mathematics and astronomy, he had put his whole heart and mind into their study. "I pondered on this subject with the whole energy of my mind," wrote Kepler, "and there were three things above all for which I sought the causes as to why it was this way and not another—the number, the dimensions, and the motions of the orbs." There were six planets: Why not more or fewer? The planets orbited at certain relative distances from the Sun: Why those distances and not others? Each planet moved at a certain speed and seemed to change its speed in a certain way: Why this particular speed and this particular change? Like many great scientific discoverers, Kepler asked simple, naive questions that most scholars of his time thought not worth asking and to which they would have responded at best with a tolerant smile for a poor schoolteacher. Part of Kepler's genius was that these questions *nagged* him.

Twentieth- and twenty-first-century scientists regard it as their mandate to try to discover why things are as they are, rather than simply to describe how they are, but that was not the case for astronomers prior to the late sixteenth century. Although it would be incorrect to say that scholars such as Ptolemy and Copernicus never pondered such causal questions, their primary concern was to describe and predict where heavenly bodies were positioned and the patterns of their movements, not to answer what *caused* them to be where they were and to move in certain patterns and at certain speeds and distances.

There was good philosophical precedent for concentrating on the one and not the other. In the fourth century B.C., Aristotle had defined a difference between mathematics (including astronomy) on the one hand and "physics" on the other. His definition could be interpreted to mean that those who studied physics were obligated to think in terms of Aristotelian "causes," while mathematicians and astronomers could ignore these concerns. Being let off that particular philosophical hook proved a great boon to astronomy in eras when looking for causes could have been no more than guesswork.

Ignoring causal questions became a pleasant habit. Medieval astronomers and philosophers thought that if one *had to* look for causes, the simple "naturalness" of the cosmos was reason enough for things being as they found them.

Two thousand years after Aristotle, Kepler bucked this tradition, thinking about such questions as, What lies behind this? According to what larger plan is this so? Why has God chosen to construct the solar system in this way and not another? Kepler knew that many of these questions might turn out to be unanswerable, but by the time he drew his fateful diagram on the board, he had begun to focus them in two questions that he thought he could answer: What line of reasoning was God using when he made things this way? and, What are the physical reasons why the universe operates as it does? He had begun to focus that second question in a way that would prove enormously important to him, asking whether one body in the solar system influences the way the others move. Maybe, for instance, the Sun did more than simply sit in the center of a neat arrangement. Kepler was not the first to wonder whether there were physical explanations for celestial phenomena, but he was the first to insist there must be and to insist on seeking them out.

When he plotted the Great Conjunctions for his students in July 1595, Kepler had already tried out and discarded some possible answers to his question about God's line of reasoning. "Almost the whole summer was lost with this agonizing labor," he reported. He had speculated, for example, that the orderly progression that underlay the relative distances of the planets from the Sun was a scheme in which Venus's orbit was twice the size of Mercury's, Earth's twice the size of Venus's, Mars's twice the size of Earth's, and so on. But neither that nor any similar set of relationships had fit. Kepler had speculated that there might be another planet, too small for us to see, between Mercury and Venus, and another between Venus and Earth, and so on, but that had not fit either.

The exercise Kepler had set himself was like some problems on modern standardized tests: Given a list of numbers, discern what mathematical regularity generates the sequence. Find the pattern that lies behind it. Break the code. On a sophisticated standardized test, one of the possible choices of answers is likely to be, "There is no pattern to this sequence, no code." In his attempt to decipher the solar system, Kepler rejected entirely the possibility of that answer. His Philippist education and his own natural inclinations caused him to believe that a universe created by God could not be random and meaningless or subject to arbitrary whim. Underlying all the seemingly disconnected aspects of nature, the complexity and the confusion, there had to be pattern, logic, and harmony. That conviction implied also that there must be hidden connections between things that seemed unrelated. Geometry, music, medicine, and astronomy had to be linked at some deep level. Kepler thought this must surely be the way God created the universe; therefore, a man created in the image of God could comprehend the logic and discover the links, with effort.

There are those who argue that Kepler's preoccupation with the harmony of the universe made him a medieval mystical throwback. He was not. His assumption of underlying harmony has become one of the pillars of the scientific method. There are indeed many connections of the sort Kepler was seeking, and these are understood now as he could not understand them. Some of the connections he experimented with turn out with hindsight to look ludicrous today, but the marvel of the man was that he thought to put them to rigorous testing. His error, and it cannot be called an error in the context of what he knew and could know in the sixteenth and seventeenth centuries, was that he had no idea how deeply such harmony lies hidden.

There are also those who would say Kepler naively contradicted himself by believing both in divine providence and in a universe not subject to the arbitrary decisions of God. But Kepler was not a naive man. He could not dismiss either side of that "contradiction" without being intellectually dishonest. Over the years, as his understanding increased, he continued, perhaps aided by his strong conviction that hidden, deeper resolutions lay behind apparent contradictions, but also out of the simple need to live with what his science and his life experience told him was true, to be exuberantly enthusiastic about both beliefs.

It was with an outburst of this exuberance that Kepler reported the Great Conjunction insight that finally did look as though it might break the code of the planetary system: "Finally I came close to the true facts on a quite unimportant occasion. I believe Divine Providence arranged matters in such a way that what I could not obtain with all my efforts was given to me. I believe all the more that this is so as I have always prayed to God that He should make my plan succeed, if what Copernicus had said was the truth." Kepler stepped back from the diagram and saw that the smaller circle was half as large as the larger circle, and that this relationship was dictated by the triangle. Was it coincidence, he wondered, that the orbit of Jupiter was about half as large as the orbit of Saturn? Could a triangle have something to do with that relationship? Saturn and Jupiter were the two planets in conjunction and the two outermost planets, and the triangle was the first figure in geometry. (The smallest number of lines from which one can create a closed geometric figure is three.) Kepler immediately began experimenting to see whether a square (the second figure of geometry) would similarly fit between the orbits of Jupiter and Mars, a pentagon between the orbits of Mars and Earth, a hexagon between the orbits of Earth and Venus, and so forth. Unfortunately this scheme did not match the observed distances between the planetary orbits. Kepler wondered whether those "known" distances were really correct.

Something else troubled Kepler about his scheme: It was too loose, leaving too much room for arbitrary choice. Beginning with a triangle and adding sides of equal length produces a square, a pentagon, a hexagon, a heptagon, an octagon, and so forth. One can go on forever adding yet another side of equal length and produce an infinite number of these so-called polygons. Certainly it would not be surprising if among those infinite polygons it were possible to find five that fit snugly between the orbits of the planets. However, to Kepler's mind, this achieved nothing, because one still had to ask why *these* polygons and not others had been chosen for the design, and why there were only six planets.

Kepler nevertheless felt he was breathing down the neck of the answer. If only he could discover why certain polygons and not others—five of them and not more—had been chosen to dictate the distances between the orbits. It occurred to him that while drawings on the chalkboard were of necessity flat (two-dimensional drawings), the real universe was three-dimensional. Perhaps it was not appropriate to apply polygons, which are two-dimensional figures, to a three-dimensional system. He considered using solid figures—three-dimensional forms—instead. "And behold, dear reader," he wrote, "you have my discovery in your hands." Kepler knew that although it is possible to create an infinite number of two-dimensional shapes in which all the edges have the same length—triangle, square, pentagon, hexagon, etc.—there is no such extensive a collection with solid, three-dimensional shapes. Experimenting with polyhedrons—solids in which all the edges are the same length, in which all the sides are the same shape, and that have other characteristics that appealed to Kepler—reveals that only five have all the defining characteristics of "perfect solids," also known as Pythagorean or Platonic solids. Nature, God, Creation, mathematical logic—they allow these five and no others.

It seemed significant to Kepler that each Platonic solid can be nested inside a sphere so that every corner of the solid touches the inside surface of the sphere. And a sphere can be nested inside any Platonic solid so that the sphere touches the center of every face of the solid. To Kepler's mind this meant there was something deeply "sphere-like" about these solids. Only of these five polyhedrons

Figure 12.2: The five Platonic solids: The tetrahedron has four faces, all of them identical equilateral triangles. The cube has six identical square faces. The octahedron has eight faces, all identical equilateral triangles. The dodecahedron has twelve identical pentagonal faces. The icosahedron has twenty faces, all identical equilateral triangles.

could it be said that each can be "inscribed into a sphere" and "circumscribed around a sphere."

Five figures thus stood apart from all other possible solid figures because of their simplicity, their mathematical beauty and perfection. Here, thought Kepler, was what God must have been thinking when he set the Sun and planets in their places. The reason there were six planets—no more, no less—was because there were five perfect solids to dictate their relative distances. As Kepler would write in the introduction to his book on the subject:

> Behold, reader, the invention and whole substance of this little book! In memory of the event, I am writing down for you the sentence in the words from that moment of conception: The Earth's orbit is the measure of all things; circumscribe around it a dodecahedron [twelve-sided regular solid], and the circle containing this will be Mars [Mars's sphere]; circumscribe around Mars a tetrahedron [four-sided solid], and the circle containing this will be Jupiter; circumscribe around Jupiter a cube, and the circle containing this will be Saturn. Now inscribe within the earth [within Earth's sphere] an icosahedron [twenty-sided solid], and the circle contained in it will be Venus; inscribe within Venus an octahedron [eight-sided solid], and the circle contained in it will be Mercury. You now have the reason for the number of planets.

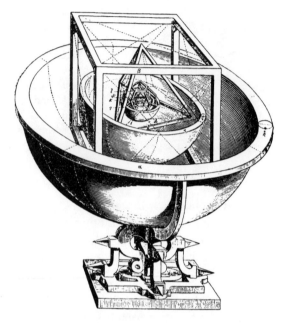

Figure 12.3: A drawing of the arrangement of the planetary orbits and the five Platonic solids, according to Kepler's polyhedral theory, from Kepler's *Mysterium.*

Likewise these five perfect solids seemed to dictate the distances apart these planets must orbit. In other words, just as, in the drawing Kepler had made for his class, the triangle dictated the size of one circle in relation to the other, he was now thinking that the requirement of fitting a cube between the sphere of Saturn and the sphere of Jupiter dictated how far apart those spheres must be. A tetrahedron must in turn dictate how the size of the sphere of Jupiter compares with the size of the sphere of Mars—and so forth.

Kepler proceeded to test this idea about God's geometric logic against Copernican theory and the available observational records "to see whether this idea would agree with the Copernican orbits, or if my happiness would be carried away by the wind." To his joy and awe, "within a few days everything worked, and I watched as one body after another fit precisely into its place among the planets." If only he had access to better observations, to be certain! Indeed, if only he could study the *best* observations in the world and could test his theory against them. The best observations in the world were those of Tycho Brahe.

The months that followed marked a change in the focus of Kepler's thinking. He had considered himself mainly interested in the big questions regarding the deep, underlying truths. Now, to find out whether his answers to some of those questions were correct, he had to turn his attention to mathematical minutiae. His earlier mathematical and astronomical training seemed sorely inadequate to the task—after all, he had judged it inadequate even for teaching school—and he realized that stupendous mathematical obstacles lay ahead of him if he was to put to rest the disturbing doubts that followed almost immediately on his elation about what he had found. In August 1595 he wrote to Mästlin for advice and help. In that letter he first called his new idea his "polyhedral theory." Mästlin's replies were cautious but full of approval and excitement.

In October Kepler reported to Mästlin that he had decided to write a book. It was clear now why God had interrupted his theological studies and sent him into what seemed such a meaningless exile in Graz. "Just as I pledged myself to God," he told Mästlin, "so my intention remains. I wished to be a theologian, and for a while I was anguished. But now, behold, God is glorified also in astronomy through my work." God, he also wrote, "wants to be known from the Book of Nature."

Kepler hoped that by the time he finished writing his book, he would be able to answer another of the questions he had been pondering: Why each of the planets took the particular length of time it did to complete an orbit of the Sun. This length of time is called a planet's period. Kepler had learned as a student that the planets nearer the Sun have shorter periods than those farther away.

The first part of the explanation for this was obvious. A planet farther from the Sun has to travel a greater distance to get all the way around its orbit, just as a runner in the outside lane of a race-track has to run farther to complete a lap than a runner in an inner lane. If all the runners in this celestial race were moving at the same speed, those farther out would take longer to complete a lap. But Kepler thought that the amount by which planets farther from the Sun lagged behind was greater than could be accounted for in this simple way. It seemed the runners in the outside lanes really were slower runners, not merely handicapped by their lane position. Pondering why this should be so, Kepler began to speculate about a possible force, resident in the Sun, that caused the planets to whirl around it. A planet closer to the Sun would feel more of the force than one farther away.

Kepler worked this idea into a formula: The increase in the length of period from planet to planet will be twice the difference of their distances from the Sun. This formula showed planetary distances that were not far off from those he had derived from his theory of the polyhedrons, but it was not correct, as Kepler himself later realized.

In every spare moment he had that autumn and in the beginning of the bitter winter of 1596, while still teaching and fulfilling the duties of district mathematician, Kepler continued to work on his book. He thought of still more questions: Was there any meaning to the particular arrangements of the polyhedrons? Was there a reason, for instance, that the cube must be the outermost, followed by the tetrahedron? Kepler was sure there had to be a reason, and he tried to discover it, while at the same time continuing to think about the force in the Sun that might be whirling the planets.

When he was putting the final polish on his manuscript, Kepler's thoughts focused more on the way a single planet moves in its orbit. Both Ptolemy's and Copernicus's models took note of the way a planet speeds up as its orbit brings it nearer the Sun and slows down as it moves farther away. It occurred to Kepler that this could be easily explained in the Copernican system by the idea that the closer the planet comes to the Sun, the more it feels the Sun's whirling force. The speeding up and slowing down were much more difficult to explain in Ptolemaic theory, with the planets orbiting the Earth. This seemed another good reason to think that Copernicus had been right about what is in the center.

Mästlin was not pleased when he heard about this idea of the force that moves the planets. He suggested it might "lead to the ruin of astronomy," as in his view Kepler was failing to respect that delicate dividing line between "physics" (which concerned itself with "causes," physical reasons why the universe operates as it does, and the nature and structure of the universe) and the use of mathematics to produce theories of planetary motion. Kepler was mixing the two areas of study by suggesting that physics (i.e., the whirling-force explanation) could explain the mathematics of a planetary system. Interestingly, Mästlin and Kepler were two of the very few scholars in Europe who should have recognized that Copernicus himself had trampled on that dividing line, from both directions, by suggesting that his mathematical theories revealed a fundamental truth about the structure of the universe—that it is Sun-centered—and that this fundamental truth made sense of his mathematical theories.

In January 1596 Kepler's studies were interrupted by the news that both his grandfathers were seriously ill, and in February he took a leave of absence from his school and traveled home to the duchy of Württemberg. Old Sebald, his father's father, in whose house Kepler had been born, died during Kepler's visit.

Kepler welcomed the opportunity to visit Tübingen to discuss his book with Mästlin in person. Such a publication, he pointed out to his mentor, would improve his stature as a scholar and make his position in Graz more secure. It seemed ironic to him that a year earlier he had been eager to leave that position as soon as possible.

Kepler also opened negotiations with a printer in Tübingen who, on the enthusiastic recommendation of Mästlin, agreed to publish Kepler's book. The only stipulation was that it be approved by the university senate. The senate, though unperturbed about approving the publication of a flagrantly pro-Copernican book, wanted two changes. First, Kepler should explain Copernicus's hypotheses and his (Kepler's) own discovery in a more understandable, popular style. Second, Kepler should remove a chapter in which he reconciled the idea of a Sun-centered universe with biblical passages that could be interpreted as supporting either Copernicus or Ptolemy. Kepler felt strongly about this chapter. He had settled in his own mind that Copernican astronomy was not incompatible with Scripture, which, he had concluded, was intended to speak to people living on Earth who had no knowledge of the true working of the cosmos. Furthermore, it was not the purpose of the Scriptures to teach them about these matters. So the Scriptures deliberately spoke in words that would make sense to such people. As he would put it later, in the introduction to another book, *Astronomia Nova*, "What wonder then if the Scripture speaks according to man's apprehension, at such time when the truth of things doth dissent from the conception of all men?" It seemed essential to Kepler that a book showing that Copernicus had been right should bring its readers along on this point. Without that, the book fell short of the glorification of God that he intended it to be. Nevertheless, he bowed to the senate's judgment that, though he might be right, interpreting Scripture was not in his bailiwick.

Kepler also visited Stuttgart during his leave, to pay his respects to the duke of Württemberg, who had earlier supported his education and then so graciously allowed him to move to Graz. This was Kepler's first experience of castle life. He had the temerity to ask for and was given a seat at the Trippeltisch, the dining table for ducal officials who were not of the highest echelon. For Kepler it was a major achievement. Gripped by (as he put it) "a childish or fateful desire to please princes" (a state of mind Tycho Brahe would have been wise to emulate at this time), Kepler presented to the duke a plan to create an elaborate model of the solar system incorporating the five solids. Whether it would be built was not settled on this occasion. Negotiations and trial runs dragged on for several years after Kepler had gone back to Graz, with him diverting much of his scarce time to providing detailed proposals, drawings, and even a paper model. At one stage, the plan was to create an enormous punch bowl. Each space between the different planetary spheres would contain a different beverage, and guests could fill their glasses from small faucets spaced around the rim, connected by means of hidden pipes and valves with the appropriate spheres. The duke finally decided to advance money to have the model fashioned in silver, but the project got bogged down in problems with the silversmith, and it was never completed.

※

WITH MUCH OF KEPLER'S TIME during his visit in Germany spent "pleasing princes" in Stuttgart, and with Tübingen proving even more hospitable—for not only Mästlin but others in the university had heard of Kepler's new idea—Kepler's absence from Graz stretched far beyond the two months he had requested. He stayed away for seven months, which almost proved disastrous to another project in which he was currently engaged and which he had left in the hands of representatives back in Graz—arranging for his marriage. The December before he had gone away, he had met Barbara Müller, the eldest daughter of a prosperous mill owner named Jobst Müller, whose estate, Mühleck, was about two hours' journey south of Graz.

It was not wise for a prospective suitor to disappear for so long. Much as Kepler's scholarly accomplishments may have impressed his former professors at Tübingen and stood him in good stead at the duke's castle in Stuttgart, they did not make him a prime candidate for husband or son-in-law in the eyes of Jobst Müller. In spite of Müller's misgivings, however, the intermediaries negotiating on Kepler's behalf had some success. In June 1596, five months after Kepler left for Germany, they urged him to return to Graz and pause only long enough in Ulm to have his and his fiancée's wedding wardrobes made "with very good silk fleece or at least the best double taffeta."

Kepler dawdled three months longer and then came back, expecting a warm welcome and congratulations all round. Instead, he learned there was to be no wedding. His prolonged absence had given Herr Müller time and cause to reconsider once again, and he was now convinced that his daughter could do better. Kepler was sorely disappointed, but he could not much blame Müller and mentioned frankly that one Stephan Speidel, who may have been working against the match for his own selfish reasons, probably only wanted to see Barbara better provided for. Though Kepler had exclaimed that when he met Barbara she had "set [his] heart on fire," theirs does not seem to have been a heated romance.

Nevertheless, that autumn, Kepler doggedly continued his suit for Barbara's hand. He may not have been ardent, but he was stubborn. The rector of his school spoke in his favor, and when, in a moment of discouragement, Kepler asked the church government either to free him from his promise to Barbara or to act on his behalf, that body also chose to influence the bride and her family to accept his proposal. Herr Müller, impressed by the authority of the church and skittish about public mockery, agreed to the union again in January 1597, and plans were set in motion for an April wedding.

As Kepler had recognized, Müller's concern for his daughter was not unreasonable. It was far from obvious that Kepler had accomplished anything of value. He had no prospect of ever being other than a poorly paid teacher, and he could promise Barbara and her young daughter by a former marriage little by way of financial support. Barbara was twice a widow. Her first husband had been a wealthy cabinetmaker and her second a district paymaster or clerk, a respected man until disreputable dealings came to light at his death. Though Kepler requested and received a pay increase on the grounds that he would no longer require lodging in the school, he wrote to

Mästlin rather pitifully as the wedding date approached, "My assets are such that if I were to die before a year is up, hardly anyone could leave worse conditions behind at his death. I must make great outlays from my own pocket, for it is the custom here to celebrate a marriage in a showy fashion."

Kepler continued his letter to Mästlin in a vein that indicates how ambivalent he was about the marriage plans. Barbara's father and Barbara herself were wealthy. The alliance gave Kepler financial security and brighter prospects. It also chained him to Graz. He wrote,

> It is certain that I am tied and fettered to this place no matter what becomes of our school. For my bride has properties, friends, and a wealthy father here. It seems that I would not, after a few years, need any salary, if that would suit me. However, I could not leave the land unless a public or private misfortune befell. A public one if the land were no longer safe for a Lutheran or if it were further pressed by the Turk . . . a personal misfortune if my wife were to die. Thus a shadow hovers over me. Yet I dare not ask more of God than He in these days allots to me.

Kepler reported that the wedding would take place under ominous constellations. The best that could be said was that the stars predicted "a more agreeable than happy marriage, in which, however, there was love and dignity."

13

DIVINE RIGHT AND
EARTHLY MACHINATION

August 1596–June 1597

THE PREVIOUS AUGUST, while Johannes Kepler stretched his stay in Germany into its sixth month, Tycho Brahe went to Roskilde Cathedral, location of the belatedly repaired chapel, for the coronation of Christian IV of Denmark. The child whose wishes Tycho had taken too lightly had come of age and was now king. Tycho had suffered recent social embarrassment over Magdalene's ill-fated betrothal, but he nevertheless made a splendid showing at the festivities. He wore the golden chains of the Order of the Elephant (a symbol that was prominent in the Chapel of the Magi) and medallions with portraits of two kings. The Brahe family was much in evidence. Tycho's brother Steen bore the royal orb. All the members of the Rigsraad held the crown to place it on Christian's head, symbolizing that the highest power lay not with the king but with the aristocratic oligarchy.

In truth, the old symbolism had almost run its course. A new age was dawning, and it would not benefit Tycho's relatives or other great noble families. The coronation oration, delivered by Bishop Peter Winstrup, celebrated a philosophy of government favored by the new king and his closest advisers, that kings rule not by the election or consent of any oligarchy but by divine right. So far that could be dismissed as only coronation rhetoric. In the days following the coronation Uraniborg seemed like its old magnificent self again and likely to endure forever, as numerous foreign guests who had come for the festivities visited Tycho. He played the magnanimous noble host and showed off the splendors of his home. Nevertheless, he and his relatives were watching the new regime with trepidation, braced for possible trouble ahead.

Scarcely a month after the coronation, the ax began to fall. Christian, conflicting philosophies of government notwithstanding, did not have complete power that could be exercised without consulting the Rigsraad, but one thing that *was* in his power was the transfer of fiefs from one noble to another. As part of a general reorganization, he took the Norwegian fief of Nordfjord from Tycho and gave it to the lord-lieutenant of Bergen.

Several times in the past that same fief had passed out of Tycho's hands and then been restored to him when he petitioned the crown. He did so now, using the opportunity to send along a summary of his accomplishments at Uraniborg, a copy of his published astronomical correspondence, and a pamphlet with woodcuts of his instruments. Tycho's petition spoke of the unfailing support King Frederick II had given this work and mentioned that the old king had intended to endow Uraniborg as a permanent research institute but was prevented by his death from doing so. Tycho even enclosed a copy of the declaration signed by the Rigsraad promising to advise Christian when he came of age to carry out his father's will, placing Uraniborg eventually under the leadership of Tycho's descendants.

In spite of the unsettled political climate, Tycho had good reason to hope that Christian would honor the old king's wishes and endow Uraniborg permanently. The chancellor to whom Tycho wrote had paid a pleasant visit to Hven, and his wife was a distant relative. Christian had clearly fallen in love with Uraniborg during his childhood visit. Tycho had, finally, repaired the chapel roof.

However, the nineteen-year-old king was strong-willed, eager to exercise his own divine right, and in no mood to respect an expensive promise made so long ago by a father he had hardly known. He chose to regard Uraniborg as a relic of the past, run by an aging aristocrat who had become too proud and powerful to respond promptly to royal commands, who had too long treated regents and rulers as his equals, who had forgotten how to be adequately deferential.

Christian's shattering reply came to Tycho through the chancellor in January 1597: Tycho had to surrender the fief of Nordfjord, and that surrender could not be postponed. The king also did not choose to endow Uraniborg permanently. Though the letter ended politely with the promise that future requests would be received with pleasure, its message was clear. All Tycho's efforts over many years to secure the future of Uraniborg and his children had come to naught. The documents and assurances he had collected were as worthless as his peasants' old claim to own their land on Hven. It counted for nothing with Christian that Tycho had fulfilled his promise to Frederick to bring glory to Denmark, so that people of other nations would come there to "see and learn that which they could hardly acquire knowledge of in any other place." Christian's father had treated Tycho like a well-loved younger brother. The son regarded him as a wearisome elderly petitioner.

Tycho was not alone in his chagrin. The king and his ministers, to weaken the Rigsraad, used the transfer of fiefs to create animosities among its members and erode the power of the noble families. Tycho's cousin lost Kronborg Castle and was moved to a castle on the fringe of the kingdom. Erik Lange lost Bygholm, which had already fallen to ruin because of his insatiable alchemical lust. Tycho's brother Steen lost Munkeliv Abbey, St. Hans Cloister, and Saebygård, and his income and influence were severely reduced.

On Hven, nothing was as it had been. Two fires broke out in the house, though they were extinguished before doing great damage. The peasants were even more restless and uncooperative than usual. Tycho was short of assistants. A set of expensive medallions with his portrait and the Brahe arms and motto that he had commissioned and intended as gifts to friends turned out to be of inferior quality—a seemingly small matter, but they had been part of his plan to restore his honor.

There had been a time when he could have quickly put minor setbacks behind him, but Tycho was an older man now—not so energetic and resilient, more defensive, overwhelmed. As the weeks passed, it seemed increasingly that everything was falling apart, and he was too tired to pick up all the pieces. He became more and more irritable and short on patience. Even so, with his income and energies severely reduced, Tycho was unwilling to take the sensible move of cutting back on research and publishing projects. If the goals he had fought for all his life looked less likely ever to be achieved, that was all the more reason not to let anything drop.

By late winter Tycho's political and financial position was deteriorating rapidly, and it became abundantly clear that time was running out for his work at Uraniborg and Stjerneborg. He put Longomontanus in charge of rushing the star catalog to completion, expanding it hurriedly from 777 to 1,000 stars and inscribing the positions of all those stars on the great globe. The work was, of necessity, not up to Tycho's usual standard of precision and verification. Other assistants had the task of taking an inventory and listing all the books in his library. The immediate plan was to close Uraniborg, move to his mansion in Copenhagen, and set up observatory, laboratory, and printing press there. Tycho used as an excuse the need to be on the spot while the diet of the nobility considered his breach-of-promise case against Gellius, but he suspected there would be no return.

The last observation recorded on Hven was on March 15, 1597, the date on which Tycho's annual pension from the crown was discontinued. His assistants had completed listing the books. After that note, the journal, which had recorded events on Hven and faithfully noted daily weather observations with no break since 1582, fell silent.

Before the move could be completed, a royal patent was issued ordering two royal commissioners, one of whom was Tycho's brother

Axel, to investigate several complaints against Tycho: that he had mistreated his villagers, allowed the pastor of St. Ibb's to violate the church ordinance of 1539, and committed other unnamed injustices.

The two commissioners found the island in a state of confusion and bustling with activity when they arrived there on April 10. The move to Copenhagen was in progress and did not appear to be temporary. Tycho had packed up everything movable, including the books, his laboratory equipment, printing presses, the great globe, and all the other instruments except the four great Stjerneborg instruments, which could not be disassembled quickly. The packing continued even while Christian Friis and Axel Brahe were meeting the bailiff, alderman, and others assessing the state of Tycho Brahe's affairs.

April 11 was clear and fair, and it would have been a fine night for viewing the skies, but by dusk Tycho and Kirsten and their six children, along with his assistants, the housekeeper, cook, butler, maids, and other servants, had turned their backs on the exquisite house, the Renaissance gardens just coming into bud, and the observatory of Stjerneborg with its silent instruments shut beneath the wooden covers. The company made their way by carriage, cart, and foot down to the little harbor and sailed away from Hven. Tycho would never again see the house he had designed to reflect the harmony of music or set foot on the soil of the beloved island that had been his home for twenty-one years.

❀

IN GRAZ, later that same month, on April 27, Johannes Kepler and Barbara Müller were wed in a splendid celebration at Barbara's own residence. In spite of Kepler's less than sanguine mood as the date drew near, the days and weeks that followed were happy for them. Barbara was only twenty-three, two years younger than he. A miniature portrait made at the time (see color plate section) shows her looking somewhat older than her age, with lovely, intelligent eyes, a sweet mouth, and a prominent nose. Contemporary descriptions called her pretty and plump. Kepler had grown extremely fond of Barbara's seven-year-old daughter Regina and treated her as his own child. Barbara was soon expecting another baby.

Not long after the wedding, the first copies of Kepler's book arrived from the printers. It was a slim volume with a long title: *Prodromus Dissertationum Cosmographicarum, Continens Mysterium Cosmographicum, de Admirabili Proportione Orbium Coelestium, deque Causis Coelorum Numeri, Magnitudinis, Motuumque Periodicorum Genuinis & Proprijs, Demonstratum, per Quinque Regularia Corpora Geometrica* (The Introduction to the Cosmographical Essays, Containing the Cosmographical Mystery of the Marvelous Proportion of the Celestial Spheres, and of the True and Particular Causes of the Number, Size, and Periodic Motions of the Heavens, Demonstrated by Means of the Five Regular Geometric Bodies). For convenience, that title is usually abbreviated to *Mysterium Cosmographicum,* or simply *Mysterium.* Looking back from old age, Kepler commented that this small book was the point of departure for the path his life would take from that time on.* He might justifiably have said the same with regard to its watershed significance for all of science, for though the polyhedral theory was erroneous, Kepler had been the first, and would be the only, scientist until René Descartes (in the 1630s and '40s) to insist on physical explanations for celestial phenomena. In the words of Owen Gingerich, "Seldom in history has so wrong a book been so seminal in directing the future course of science."

Kepler hastened to send copies to other scholars, requesting their opinions. Galileo Galilei, then teaching at the University of Padua, was not yet well known and perhaps not known at all to Kepler, and *Mysterium* probably came into his hands purely by serendipity through a third person. But Galileo wrote to Kepler that though he had read only the preface so far, he was looking forward with pleasure to reading

---

*Science historian Bruce Stephenson has quipped that "most of the larger problems that concerned Kepler throughout his career were raised in this book—raised, indeed, in its title!"

the rest. He also mentioned that he had been a Copernican for some years but not admitted it publicly for fear of the ridicule of his colleagues. In a return letter written in his most exuberant style, Kepler urged Galileo to espouse Copernicanism openly, for "would it not be better to pull the rolling wagon to its destination with united effort?" He also begged for Galileo's opinion of *Mysterium:* "You can believe me, I prefer a criticism even if sharp from a single intelligent man to the ill-considered approval of the great masses." Galileo did not reply. There would be no further correspondence between them for thirteen years.

Among other scholars to whom Kepler sent copies of his book, the reception was mixed. Mästlin agreed with him entirely. Johannes Praetorius, a professor from Altdorf who responded favorably at first, changed his mind on closer reading and declared that astronomy "could derive no profit from these speculations. The planets' distances should be found by observation"; they meant nothing beyond that. Professor Georg Limnäus from Jena was overjoyed that someone was "reviving the Platonic art of philosophizing."

In his correspondence with Limnäus, Kepler requested information concerning a famous Danish astronomer, Tycho Brahe. Limnäus's reply mentioned that Nicolaus Reimers Bär was a "specialist" who had "spent some time with" Brahe. Fatefully, Limnäus failed to add that there was antagonism between Tycho and Bär.

Bär, or "Ursus," as he had latinized his name (*ursus* is Latin for *Bär,* or in English, bear), had risen dramatically in the world since his visit to Hven. Using false credentials, he had contrived to ingratiate himself with Rudolph II, emperor of the Holy Roman Empire, who was always eagerly on the lookout for a good astrologer. So successful had Ursus's ploy been that he was now ensconced as imperial mathematician at Rudolph's court in Prague. In spite of his status, Ursus was not one of the scholars to whom Kepler originally sent copies of *Mysterium,* but Ursus noticed the book listed in the Frankfurt catalog and wrote to Kepler requesting a copy.

That request was not Kepler's first contact with Ursus. A year and a half earlier, in November 1595, at the urging of a supervisor in Graz who praised Ursus highly, Kepler had written to tell him about his polyhedral theory. It was not characteristic of Kepler to be dishonest in his dealings, but on this occasion, in a flourish of ill-considered disingenuity, Kepler had written, "The little knowledge I have in astronomy I acquired with you, that is, with your books, as my teacher." In fact, Kepler had never read Ursus's books. His letter also declared, "I love your hypotheses," and he closed with the words, "Take care of yourself, for the sake of the stars and our science, O Pride of Germany!" signing himself, "Your excellency's pupil." Kepler would live to regret that hyperbole. Ursus's hypotheses—with which Kepler certainly did not agree (he admitted later in the letter), no matter how much he "loved" them—were those that had the Sun orbiting a motionless Earth and the other planets orbiting the Sun. In other words, Tycho Brahe's system.

Ursus had not troubled to reply to Kepler's letter, but he had not discarded it either. Learning about *Mysterium,* he recognized a delicious opportunity: This obsequious young fool, this Kepler, was no longer a nobody. He had authored a book. Ursus saw that he could strengthen his claim that the "Tychonic system" was his invention by reprinting Kepler's adulatory letter in his own forthcoming book *De Astronomicis Hypothesibus* (On Astronomical Hypotheses), a vitriolic attack on Tycho. To scholars who read the book, it would appear that Kepler had entered the contest on Ursus's side.

Kepler had no way of knowing why in May 1597, a year and a half after he had first written Ursus, Ursus was suddenly so friendly, addressing Kepler as "most distinguished man" and "esteemed friend." Kepler innocently sent Ursus not one but two copies of *Mysterium,* requesting that he pass one on, if the opportunity arose, to Tycho Brahe.

IN THIS SAME late spring of 1597, Tycho was facing a fresh set of problems in Copenhagen. Though by and large Uraniborg and the University of Copenhagen had maintained a relationship that was useful to both—with the university sending some of its most promising students, such as Longomontanus, to Uraniborg to take advantage of the opportunities there—Tycho had some jealous enemies among the university faculty. Now his presence in the city, in a mansion with an observatory that no university could match, provoked afresh the resentment of men who argued that Tycho's research drained the university of financial support and its ablest students.

The animosity came not only from astronomers but also from theologians who were pleased to see him at bay. Tycho found himself in the crossfire between two warring schools of Lutheran theology, the Philippists, who were enthusiastic about the study and advancement of science, and the Gnesio-Lutherans, who were less so. There were other differences, including their attitude toward marriage. Particularly relevant for Tycho and Kirsten, the Philippists tolerated mutual pledges of betrothal that did not involve a church wedding, while the Gnesio-Lutherans strongly supported the royal ordinance of 1582 that had already caused Tycho to stop taking Communion.

Leaving the island also did not mean leaving behind the problem of Tycho's tenants. The report of the two royal commissioners led to a summons to court. Tycho and the peasants of Hven were to appear before the king himself. This time the peasants' charges had more to do with Tycho's relationship with the church of St. Ibb's than with their own oppression. Perhaps they were aware that maintenance of church property had been an issue between Tycho and the king before, and that there were also problems between Tycho and theologians at the university. The villagers accused Tycho of letting the church deteriorate and pocketing incomes and tithes for himself, of expropriating glebe lands, tearing down parsonage buildings, underpaying the pastor, and appointing and dismissing pastors at whim. Exorcism had been omitted from the ritual of baptism at St. Ibb's, and Tycho had not corrected this omission. Tycho countercharged that the peasants had maliciously damaged the Stjerneborg instruments. The proceedings were discontinued pending further investigation.

The pastor of St. Ibb's, Jens Jenson Wensøsil, fared worse in this round of the investigation than Tycho did. Wensøsil was charged and found guilty of omitting exorcism from the ritual of baptism and failing to punish and admonish Tycho Brahe for living a sinful life with his common-law wife and for missing Communion for eighteen years. The attack on Wensøsil was clearly a thinly veiled attack on Tycho. It was Tycho who had ordered Wensøsil to omit exorcism, thereby revealing his own Philippist rather than Gnesio-Lutheran sympathies. It was Tycho who had chosen not to take Communion and violate the ban having to do with common-law marriages. Crown lawyers knew that attacks on Tycho's marriage were pointless, for it was legal under the ancient Jutish law to which any nobleman had the right to appeal. Unable to touch Tycho himself, his enemies had brought down the vulnerable pastor instead. Wensøsil was imprisoned in a dungeon for a month and, according to Tycho, "would have been beheaded if powerful friends had not intervened." When Wensøsil emerged, he fled to Tycho's mansion.

Tycho suffered an even greater indignity than having to defend himself against his peasants before the king or see his pastor take the brunt of the hatred that was intended for him. The town constable came to Tycho's door in the name of the king, who had a view of Tycho's mansion from Copenhagen Castle across the water, and ordered Tycho to remove the instruments he had mounted on the bastions and the city walls behind the mansion. Christian claimed they spoiled his view. All work ended at Tycho's Copenhagen observatory. He became, again, a laughingstock at court.

On June 1 Tycho sent Longomontanus away with a letter of recommendation to future employers. Franz Tengnagel, a young Westphalian nobleman who had been one of Tycho's assistants since 1595, departed for the Netherlands. The next day, Tycho and his household, still more than twenty people, left Copenhagen.

The long, lumbering column of carriages and heavy wagons took the road south. The carriages carried Tycho and Kirsten, their six children (all but Magdalene were in their teens), and the pastor Wensøsil. The wagons were loaded down with Tycho's three thousand books, his manuscripts, his laboratory equipment, the printing press, all his astronomical instruments except four in Stjerneborg, household goods, furniture, the family's clothing and personal belongings, and iron-bound chests containing all their wealth in gold, silver, and jewels. Household servants rode in some of the wagons, and there were armed men on horseback for protection, as well as extra horses and numerous teamsters for hauling, loading, tending the draft animals, prodding them on and off the boats where there was water to cross, wading with them through river fords, and prising wagons out of muddy ditches. Tycho probably also brought along good horses for himself and his two sons, so that they could make periodic checks on the entire length of the caravan.

Tycho was taking his entire life with him, except for two empty mansions and four instruments in an observatory. He had always been somewhat prepared for this contingency, despite outward confidence and a recent dangerous tendency to forget how ephemeral the favor of princes could be. When he ordered his first quadrant at Augsburg, he had had it designed to be easily dismantled, and he had followed that practice ever since except with the four largest instruments. "An astronomer must be cosmopolitan, because ignorant statesmen cannot be expected to value their services," had been his youthful, prophetic words. More recently, while he had seemed to be struggling to continue at Hven, he had also been making thorough and extensive preparations to make this move and do it in a style befitting a man of his wealth and stature.

The caravan wound past Vordingborg, the massive castle fortress where Tycho had lived as a boy. There, all the wagons and the carriages were loaded onto ferries for the next leg of the journey to the port of Gedser on the southern tip of Denmark, where they took a ship for

Rostock, less than two months after saying farewell to Uraniborg and Hven. Tycho, at fifty years old, was an exile from the country that had celebrated him and supported him and his work for twenty-one years.

Uraniborg stood deserted, an odd, magnificent shell of a house. On the wall where the mural quadrant had been, Tycho's portrait gazed at nothing. The library shelves were empty, the great globe gone, the alchemical furnaces cold. No water spouted from the fountain at the center of the house. No summons came to the garret rooms through the secret communication system. Beyond the formal gardens, the four great instruments of Stjerneborg gathered dust in the darkness, the wooden roofs that protected them from the elements stayed in place night after night, never pulled back so that the sights could be pointed at the stars. David Pedersen, Tycho's bailiff, summoned laborers from Tuna each day to maintain the manor and occasionally received aid from representatives of the crown to maintain order, for the peasants of Hven were ready to tear house, gardens, and observatory down stone by stone, use the materials elsewhere, and make the land a pasture again.

Eventually, that was what they did. Though all over Denmark and Sweden castles and stately homes older than Uraniborg are still inhabited or lovingly preserved, nothing remains of Tycho's. There is no record of a protest or any regrets when the cornerstone of his paper mill was found on Hven and removed to Knutstorp in 1824. It was donated then to the historical museum in Lund, but when the museum showed little enthusiasm about displaying it, it was trundled back to Knutstorp.*

Tycho's name remained anathema on Hven for generations. Not until the late twentieth century did attitudes on the island change and the legend of the evil lord who had built his detestable palace on the common land finally fade. Conservators of the site of Uraniborg planted a hedge to show its footprint, restored part of the outer wall

---

*The paper-mill cornerstone stands today in a place of honor just outside the door of Knutstorps Borg, its inscription still legible.

The statue of Tycho Brahe that stands in the restored portion of the garden at the Uraniborg site on Hven.

and a quarter of the gardens, and made Stjerneborg a well-marked archaeological site. They built a small museum and are currently also trying to repair Tycho's reputation, for the image of him as a half-mad, fire-breathing tyrant had spread into the history books. Revisionist descriptions of Tycho give him much more sympathetic treatment than he has received in the past. In the restored garden at Uraniborg, a splendid statue of him, robed, instrument in hand, gazes with powerful intensity at the heavens.

In December 1597 another Mars opposition brought Mars particularly close to Earth. "I wish he had been there," Kepler would write much later, "because this opposition was a marvelous opportunity, not often recurring within a man's lifetime, for finding Mars's parallax." The parallax of Mars would not be a weapon in the winning of the Copernican revolution. Tycho's Mars observations, however, would be.

14

CONVERGING PATHS

June 1597–November 1598

TYCHO HAD RESIGNED HIMSELF to getting little work done during the move into exile, but traveling the alternately dusty and muddy roads did give him time to think about the future. Though he may have suspected that he would never go back, Tycho had not given up hope. He was not the first among his powerful family to be driven out of Denmark, nor would he, if he could manage it, be the first to return and regain former status. His foster uncle Peder Oxe and his brother Knud had managed no less.

Tycho and his entourage settled into lodgings in Rostock in mid-June 1597, and the busy academic city was a tonic to his bruised ego. He had lived here for a while as a student and left missing most of his nose. There were still many Danes and good friends at the university, and they received Tycho warmly, treating him as one who came in triumph and with great honor, not in disgrace. His prodigious intellectual and technical accomplishments and the splendors of Uraniborg were legendary here. In Rostock Tycho and his family could enter St. Mary's Church and receive Holy Eucharist together, something they had not been able to do in Denmark for eighteen years.

Tycho began to put into motion the plans he had made during the

journey. He had decided on several points of attack. The campaign
to restore his honor and position in Denmark would begin immediately. It would include a direct appeal to King Christian, with pressure brought to bear simultaneously from within Denmark (from
Tycho's still powerful relatives) and from influential people abroad.
At the same time, he would settle his family, entourage, and equipment in a semipermanent location and resume his astronomy and
the publication of his books. The quicker he could proceed with
that, the more secure the future would be for them all, whether in
Denmark or somewhere else. The third part of his plan was to exploit
his network to secure a new patron among the royalty of Europe—
someone else with fiefdoms and islands to bestow.

Tycho drafted his appeal to King Christian, explaining why he
had moved abroad without taking leave of the king, and reminding
Christian of the promises made by his father Frederick, the Regency
Council, and the Rigsraad. Rather than abide by those promises,
Christian had cut off Tycho's income. Tycho closed his letter by insisting that he would rather serve Christian than any other master
but would seek a patron elsewhere if such service could not be rendered on "reasonable terms, and without damage to me."

Tycho had never been an obsequious courtier, but he had known
when and how to appear deferential and how to flatter kings. Yet his
appeal to Christian had the tone of a letter between equals who had
the right to scold each other about unfulfilled promises, and it was
certain to bring a negative response. Either Tycho wanted Christian
to turn him down, or his pride and anger got the better of his good
sense and political savvy.

Perhaps Tycho was more resigned to leaving Denmark than he
willingly admitted at the time. Uraniborg had become an enormous
administrative burden. The research for which he had had such high
hopes and designed his finest instruments—the search for Mars's
parallax—had ended in disappointment. Ahead were new challenges, and in Rostock he was filled with fresh energy and a renewed

sense of his own worth. He could not in good conscience fail to make an appeal to Christian and apply what pressure he could on the king. However, he also used that appeal to remind Christian, and himself, that the king was not dealing with a groveling underling but with a proud man of enormous intellectual and social stature, who was willing to return only on his own terms. In Tycho's domain—in the intellectual world, in astronomy—he clearly felt he ruled as surely by divine right as Christian ruled Denmark. With similar impatient, well-warranted, but ill-advised arrogance, Galileo would later incur the wrath of Pope Urban VIII. Tycho had at least been astute enough to remove himself and most of his worldly and scholarly treasure out of Christian's reach.

While waiting for Christian's reply, Tycho began putting to use the European network he had cultivated for years. Duke Ulrich of Mecklenburg, King Christian's grandfather no less, agreed to intercede with the king. His intervention may have done more harm than good, for he praised Tycho as a man whose equal it would be difficult to find and who was famous in many lands. Christian, wishing for a subservient, compliant Tycho, surely did not take well to this reminder. Tycho sealed his friendship with Duke Ulrich by lending the guardians of the duke's two nephews ten thousand dalers—which also did not escape Christian's notice. Tycho had cried poverty in his letter of appeal.

In early September there had been no reply from Christian, and Tycho sent out a feeler in the campaign to find a new patron. He still thought in terms of islands, and the first letter went to Lord Chancellor Erik Sparre in Poland, asking about the possibility of obtaining an island in the Baltic from King Sigismund. For political reasons having nothing to do with Tycho, that suggestion came to naught.

Also in early September, Tycho and his entourage left Rostock and took to the roads again. Tycho had decided to seek the hospitality and counsel of Viceroy Heinrich Rantzau of Schleswig-Holstein.

Learning that the viceroy was not at home at his seat at Segebert Fortress, they traveled on to find him in Bramstedt.

Rantzau was a prodigious and highly respected scholar, older, far richer, and more powerful than Tycho, and equally famous. He and Tycho had much in common. Rantzau had built many palaces, as richly adorned as Uraniborg with Latin epigrams, pavilions, and pyramids. Now in his seventies, he was still an astute politician. The two men proceeded to pool their knowledge and experience of the politics of royal patronage. Though Rantzau lived like a prince, he was only a viceroy and not a suitable patron for Tycho. However, he was in a position to lend Tycho a castle where he and his household could live until the moment came to return to Denmark or move elsewhere. Rantzau had plenty of castles. Tycho chose Wandsburg, on the outskirts of Hamburg.

In late September Tycho's coach, drawn by six horses and followed by the long train of slower conveyances, rattled over the drawbridge into the courtyards of this massive Renaissance palace. It was a splendid dwelling, worthy of a man of Tycho's stature, where he could maintain the princely image he thought necessary for regaining the favor of King Christian or winning a new patron. It was also near the city, where there were engravers and printers for the publications Tycho was planning, and it had a tower with a clear view of the skies, roomy enough to set up instruments.

In mid-October a courier from the Danish court picked his way among servants unpacking instruments, printing press, furniture, and personal belongings to hand a letter to Tycho's secretary, who broke the seal and read it aloud. It was Christian's reply, and it was openly hostile. He had written that he took great offense at the tone of Tycho's appeal, which Tycho had composed "audaciously and not without great lack of understanding, as if We were to render account to you concerning why and with what cause We made changes on Our and the Crown's estates." Tycho had "not blushed to [write] as if you were Our equal. . . . from this day on, We shall be otherwise re-

A woodcut of Wandsburg Castle, dating from 1590.

spected by you if you expect to find in Us a gracious lord and king."
Christian also took offense at Tycho's going abroad with "his woman
and children" (a reference to Kirsten's low status) to beg from others,
implying before all the world that Christian and Denmark were not
wealthy enough to support them.

It was a brutal letter and clearly not drafted carelessly in a fit of
pique. It confronted Tycho's appeal point by point: Tycho had
abused his position as a nobleman by appropriating incomes and
tithes of the Hven church; he had failed to maintain the church
buildings; he had not paid the parson a suitable wage; he had permitted economic abuse of his tenants. Since all these charges had indeed been brought formally against Tycho, and the record was not
yet clear whether he was guilty, the king may have had substantial
reason for outrage at the argument that the crown had unfairly transferred fiefs from Tycho. The letter went on to insinuate that there
was reason to doubt Tycho's claim that the transfer of fiefs had so impoverished him that he had had to sell his right in Knutstorp to support his astronomy, in the interest of the honor of Denmark and the
future of science. News had reached the king that Tycho had money
"to lend in thousands of dalers to lords and princes, for the good of

your children and not for the honor of the kingdom or the promotion of science." As for Tycho's willingness to return on "reasonable terms," Christian replied, "If you would serve as a mathematicus and do what he ought to do, then you should first humbly offer your service and ask about it as a servant ought to do . . . afterwards We shall know how to declare our will." Meanwhile, Christian would not "trouble Ourselves whether you leave the country or stay in it."

Whatever ambivalence Tycho had about returning, and however much he had been expecting the worst, this reply was a blow. Living in exile and seeking a new position were no easy prospect at his age. As he expressed his misgivings in a letter to a Danish friend: "No doubt the time will come when experience and circumstances will render [Christian] more clear-eyed and sensitive . . . about what is of greater value to his realm than other useless things . . . but it will be too late for me and my researches."

Characteristically, Tycho took refuge in his self-image as one of the great men of history, many of whom, including the poet Ovid, had experienced the bitterness of exile. Tycho wrote an "Elegy to Denmark," 102 lines of Latin verse modeled on Ovid's elegies. After sending a copy of that to Rantzau, he moved quickly to other matters. Though he would not yet cease pressuring the Danish government to restore him to Hven, it was all the more urgent that Wandsburg become a temporary Uraniborg while he sought a new patron.

The day after venting his rage toward King Christian in the elegy, Tycho resumed his systematic observations of the planets and moved forward toward the publication of a splendid new book. *Astronomiae Instauratae Mechanica* would describe his instruments in words and woodcuts, document their superiority, and tell of his life, his work, and Uraniborg. Tycho saw this as a sort of extended résumé, but in a lavish, elegantly bound printing suitable to be a gift for kings—a powerful credential in Tycho's patron search. By the end of the year he was busy writing a dedication for the book to Rudolph II, emperor of the Holy Roman Empire, who ruled from his court in Prague, and whose imperial mathematician was Ursus.

Tycho planned to use his star catalog as another credential. During the last weeks on Hven he had driven himself and his assistants, trying to finish "filling out the thousand," in other words, to bring his catalog up to the thousand stars that were usually included in the ancient star catalogs. Now, at Wandsburg, the assistants who had accompanied him into exile or rejoined him (possibly only two of them at this time) labored in the tower to complete that task. Neither Tycho nor his heirs ever published the catalog, perhaps because Tycho was unable to achieve the quality he hoped for by cross-checking independent sets of data. However, though the catalog may not have been, by his standards, good enough to publish, it was good enough to impress European rulers. Tycho produced an elegant manuscript version, hand-lettered on vellum parchment.*

At New Year 1598, Tycho dedicated this volume also to Emperor Rudolph as a gift and entrusted its safe journey to Prague to his sixteen-year-old son Tycho, who was starting out on travels for his own education. In his search for a new patron, the elder Tycho had chosen his target.

Twenty-two years earlier, when he had attended Rudolph's coronation as King of the Romans in the lead-up to becoming Holy Roman Emperor, Tycho had made a friend of Thaddeus Hagecius, Rudolph's personal physician. They had continued to correspond over the years during which Hagecius had become one of Rudolph's most trusted advisors. Hagecius was one of the men to whom Tycho had written about his Mars parallax observations. Now, by letter, Tycho consulted Hagecius about whom to approach at the court in Prague and how to go about it. Tycho had additional contacts there as well, including the imperial librarian (an acquaintance since student days in Basel) and several Austrian noblemen who had visited Uraniborg. There were

*It was this catalog—not a better one—that Kepler later used.

others not powerful enough to influence imperial decisions who nevertheless were useful in keeping Tycho well informed.

As the elegant pages of his new *Mechanica* came off the press in the spring of 1598, Tycho had the books bound differently depending on where he planned to send them, some in leather, others in vellum or fine silk with metal clasps. Perspective drawings that painstakingly copied nature and drew the viewer deep into the picture were the height of artistic fashion in the late sixteenth century, and the thirty-one woodcuts and engravings in Tycho's book—quite apart from their value in documenting and showing off the wonders of his instruments—were superb examples of this "mannerist" style. Far more than a scientific treatise, *Mechanica* was intended to be the equivalent of a coffee-table book for the palaces of Europe.

Former assistants who were noblemen were soon carrying opulent copies of *Mechanica* and the star catalog to princes, bishops, archbishops, and other rulers who had valuable contacts in Denmark and Prague. Tycho's carefully chosen couriers had access to the courts of Europe and could converse comfortably with princes and other rulers about Tycho's achievements and stir them to outrage concerning his current tribulations. Franz Tengnagel, the Westphalian nobleman who at the age of nineteen had joined Tycho at Hven in 1595 and remained with him until the day before Tycho left Copenhagen, had rejoined him in exile. Though now only twenty-two years old, Tengnagel was an extraordinarily effective courier. Tycho sent him to Archbishop Elector Ernest of Cologne, whom Rantzau knew to be a particularly influential cousin of Emperor Rudolph. Not only did Tengnagel win the archbishop's deepest sympathy and generous promises of assistance for Tycho, but he came away with a gold medallion and a fine riding horse for himself as well.

The archbishop, true to his word, wrote two letters—one addressed to Rudolph, assuring him that "the whole German fatherland" would bless him if he granted generous patronage to Tycho Brahe, the "unique and most laudable restorer of the sciences"; the second to the emperor's closest adviser, Johannes Barvitius, pressing Barvitius to facilitate Tycho Brahe's case. These letters were put into Tycho's hands so that he could present them himself. Tycho, meanwhile, covered his bases by sending gift books to other parts of his network. Prince Maurice of Orange promised to try to arrange public support for Tycho in Holland.

At the same time, Tycho was rebuilding his staff. He had kept in touch with many other former assistants besides those he was using as noble couriers, and he wrote to some of these at German universities, inviting them to join him. Wandsburg was on one of the most direct routes from Denmark to the rest of Europe, and as word spread of Tycho's new address, many young Danish scholars stopped by for visits. They gave him repeated boosts for his ego and another means of keeping abreast of what was happening all over the Continent, not only in politics but also in the scholarly world.

In early March one such visitor arrived, bringing Tycho two books and a letter. One of the books raised Tycho's hackles the moment he saw the author's name: Ursus, none other than the Nicolaus Reimers Bär who had been such an obnoxious visitor at Uraniborg in 1584 and had caused Tycho so much worry and grief since. Though Ursus had never again been a presence at Tycho's table or nosing about in his library, he had made a habit of surfacing now and again, claiming Tycho's planetary system as his own and referring to Tycho as more astrologer than astronomer and not likely to achieve anything important. Since the scholarly world was well monitored by Tycho's network of former students and colleagues, none of these incidents escaped Tycho's ears.

The book that Tycho now held in his hands was even more disturbing. Not only did it attack Tycho as an astronomer and plagiarize his work, it was also a scurrilous personal assault on him and his family. According to Ursus, Tycho had been forced to leave Denmark because he had committed an atrocious crime. Ursus mocked Tycho's disfigurement, commenting that Tycho could "discern double-stars through the triple holes in his nose." Of Tycho's wife and eldest daughter, Ursus wrote "the daughter . . . was not yet nubile at the time I was there and so not of much use to me for the usual purpose. But I don't know whether or not the merry crew of friends who were with me had dealings with Tycho's concubine or his kitchen-maid." If there had previously been any question whether Ursus had actually plagiarized Tycho's system, this book made it clear he had. "Let it be theft," Ursus jeered, "but it was intellectual. Learn to safeguard your possessions hereafter." The book also brought Tycho the chilling news that Ursus was now Rudolph II's imperial mathematician. Though Tycho could never hold that title himself, because it was too lowly for a nobleman, Ursus clearly stood in his way.

As Tycho leafed through Ursus's book, he saw that Ursus had included a reprint of a letter from a young scholar in Graz. The name, Johannes Kepler, meant nothing to Tycho, but the tone of his letter—Kepler had written to Ursus, "The bright glory of thy fame . . . makes thee rank first among the *mathematici* of our time like the sun among the minor stars"—was enough to ensure eternal enmity between Tycho and this fawning young idiot.

When Tycho put aside Ursus's book and picked up the second book he had received that day, he found that by astounding coincidence it was from this same Kepler. The book introduced a new scheme for explaining the planetary orbits using the Platonic solids. The third item Tycho received, the letter, was also from Johannes Kepler, asking Tycho to give his opinion of his book. There were only two possibilities: Either this Kepler lacked the wits to foresee that Ursus's book would inevitably come to Tycho's attention, or else Ursus had published the letter without Kepler's knowledge or permission.

Tycho did not react by tossing Johannes Kepler's book into the fire. Instead he looked carefully at the little volume. Tycho disagreed with its espousal of Copernican theory, but the book gave clear evidence of a brilliant mind entirely out of sync with a mind such as Ursus's. Tycho had experience judging young talent and knew it

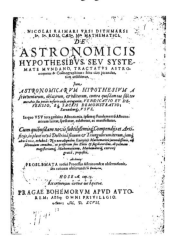

The title page of Ursus's scurrilous book *On Astronomical Hypotheses*.

when he saw it. He must also have recognized a fellow master of adulatory and not necessarily sincere rhetoric. Here, side by side, were Kepler's letter in Ursus's book that praised Ursus as "first among the *mathematici* of our time like the sun among the minor stars," and Kepler's letter that called Tycho "the prince of mathematicians not only of our time but of all times." Perhaps Tycho was wryly amused that this ranking did, in fact, place him somewhat above Ursus.

Tycho concluded that a campaign to discredit and destroy Ursus could no longer be postponed. He wrote to Longomontanus, asking him to try to remember everything he could about Ursus's visit to Uraniborg and to come to Wandsburg as soon as possible to discuss the matter. He also began to round up copies of Ursus's book to burn them. Finally, he set in motion a plan to turn Johannes Kepler's blunder to his own advantage.

Due to lack of dependable mail or courier service in late-sixteenth-century Europe, though Kepler had sent his book off to Tycho in the late winter of 1598, no news from or concerning Tycho would reach Kepler until the following late autumn. During those beleaguered

months, when Kepler was twenty-six, he had many other concerns in
addition to who was reading and reacting to his book, but he found
the time to pen a whimsical description of himself, in the third person, as a "house dog":

> That man [Kepler] has in every way a dog-like nature. His ap-
> pearance is that of a little house dog. His body is agile, wiry,
> and well-proportioned. Even his appetites were the same: he
> liked gnawing bones and dry crusts of bread, and was so greedy
> that whatever his eyes chanced on he grabbed; yet, like a dog,
> he drinks little and is content with the simplest food. His
> habits were also like a house dog. He continually sought the
> goodwill of others, was dependent on others for everything,
> ministered to their wishes, never got angry when they reproved
> him and was anxious to get back into their favor. He was con-
> stantly on the move, ferreting among the sciences, politics, and
> private affairs, including the most trivial kind; always following
> someone else, and imitating his thoughts and actions. He is
> impatient with conversation but greets visitors just like a dog;
> yet when the smallest thing is snatched away from him he flares
> up and growls. He tenaciously persecutes wrong-doers—that
> is, he barks at them. He is malicious and bites people with his
> sarcasms. He hates many people exceedingly and they avoid
> him, but his masters are fond of him. He has a dog-like horror
> of baths, tinctures, and lotions. His recklessness knows no lim-
> its; yet he takes good care of his life.

Kepler's relationship with his father-in-law was not going well.
"He hurt me with his contempt and mocking," Kepler wrote,
"though my imagination made this problem worse than it was in reality. He wanted to take away or alienate my stepdaughter. I provoked him [all the more] through the intensity of my anger."

Kepler and Barbara's first child, Heinrich, had been born on

February 2. The horoscope Kepler had cast for him promised a life far
happier than those of the two previous Heinrich Keplers, Johannes's
father and brother. The stars suggested that he would be like his father, "only better, with charm, nobility of character, nimbleness of
body and mind, and mathematical and mechanical aptitude." None
of that was to be, for the baby Heinrich lived only two months.

In June Kepler wrote to Mästlin, who had also recently lost his
own little son, "Time does not lessen my wife's grief, and this passage
[from Ecclesiastes] strikes at my heart: 'O vanity of vanities, and all
is vanity.'" In the same letter Kepler spoke of his increasingly serious
misgivings about the religious situation in Graz. The tension between Catholics and Lutherans had been worsening from year to
year and was beginning to erupt in open hostility. The present esca-
lation dated from December 1596, the December after Kepler's return from his leave in Germany. Archduke Ferdinand II had come of
age and assumed rule over Inner Austria, including Styria and Graz.
Ferdinand's father had tolerated Protestants in his lands, but his wife,
Ferdinand's mother, was a fervent Catholic who was appalled by this
tolerance, and Ferdinand himself had grown up in Catholic Bavaria
and been educated by Jesuits. Protestants feared that Ferdinand
would enforce his rights under the Peace of Augsburg and compel all
his subjects to convert to Catholicism.

For the first few months after Ferdinand's coming of age, those fears
began to seem unfounded. However, in the summer of 1598, when
Kepler brought the matter up in his letter to Mästlin, Ferdinand was
meeting Pope Clement VIII in Rome, and the citizens of Graz were
waiting with trepidation to see what changes might follow. They were
right to worry: The Counter-Reformation in Graz was about to begin
in earnest.

Kepler watched in despair as his fellow Protestants—buoyed by
false confidence because they had so long held the balance of power in
their hands—invited their own disaster, openly taunting Catholics,
circulating vulgar caricatures of the pope, even mocking the worship

of Mary with an obscene gesture from the pulpit. The Catholics re-
taliated. In the hospitals for the poor, Protestant patients were passed
by without treatment. There were new high taxes for Protestant buri-
als. Finally, the Catholic archpresbyter forbade every Lutheran sacra-
ment, including marriage and Communion.

The Lutherans appealed to Archduke Ferdinand, but on September
13 Ferdinand responded with even harsher measures, and this time
they affected the Keplers directly. The Lutheran college and all
Lutheran church and school ministries would be closed within fourteen days. Ten days later the archbishop banished all Protestant ministers and teachers. They had to be gone from the city before the week
was up, on penalty of death.

Again the Lutherans protested and summoned the assembly of the
Estates of Styria. The counselors begged the archduke to repeal his
decree. Instead, Ferdinand ordered that all collegiate preachers, rec-
tors, and school employees must not only be out of Styria within the
previous deadline, but they must depart Graz and its environs *by
nightfall.* Anyone failing to obey would face "the loss of life and
limb." Kepler and the rest of the faculty at his school hastily packed
a few essentials and fled into the country outside the city, leaving
their families behind and hoping the archduke would relent. Not
one of them other than Kepler was allowed to return.

Kepler had several advantages working in his favor. In addition to
his position at the school, he was the district mathematician. Because
this was a neutral office, neither Protestant nor Catholic, it provided
a valuable argument for allowing Kepler to remain in Graz. Kepler
also had friends in high places who could make this argument for
him. He had been carrying on a lively correspondence about scientific
questions with the Bavarian chancellor Hans Georg Herwart von
Hohenburg. Although the chancellor was a devout Catholic, he remained a helpful and sympathetic friend in the present crisis. Kepler
had other influential contacts as well and knew that Ferdinand him-

self enjoyed hearing about his scientific discoveries. When Kepler petitioned for permission to return to the city, his petition was granted. For the time being, he was safe.

Protestants still in Graz temporarily found ways to circumvent the ban on Protestant worship by attending services at nearby country estates whose Protestant clergy had not been sent away. There was soon an end to that as well, and new ordinances required Protestants to baptize their children as Catholics and marry only in Catholic ceremonies. "Heretical" books were banned, including Luther's translation of the Bible. Searches took place throughout the city, and ten thousand volumes were burned.

Kepler had no teaching duties now that the school was closed. He spent his days immersed in new thoughts about the harmony of the heavens. Von Hohenburg loaned him books he could not find in Graz, but Kepler did more than read. As he put it, "He who distinguishes himself by intellectual agility has no inclination to concern himself much with the reading of the works of others. He does not want to lose any time." He was beginning the speculation that would result in his book *Harmonice Mundi* twenty years later, and also keeping his eye out for a new job.

Thus it was that in the early autumn of 1598, Kepler, like Tycho, was seeking employment. He was making the best of a far less impressive network than Tycho's. An appointment at the University of Tübingen seemed the obvious solution, but Kepler's inquiry there aroused no interest at all. Meanwhile, the letter Tycho had written in April, reacting to Kepler's letter and book and praise of Ursus, and a letter from Mästlin concerning the same, were floating about in European mail limbo. Kepler still had not learned that Tycho had actually penned a vague invitation to join him. There was no easy way Kepler could have acted on that invitation, even had he known of it. Tycho was far away, and the offer was not concrete enough to encourage Kepler to make such a journey. He could only go on yearning in vain for a glimpse of Tycho's superior observations of the heavens.

⚜

T Y C H O , from his Wandsburg palace, continued to play his cards brilliantly during that summer and autumn of 1598, showing no signs of having lost his touch as a courtier: Emperor Rudolph's three most powerful advisers—his personal physician Hagecius, his vice chancellor Rudolph von Coraduz, and Johannes Barvitius—were all working in Tycho's cause. Rudolph heard from this undisputed inner circle at his court—as well as from his gem artist and astrologer Caspar Lehmann—that princes, archbishops, and scholars all over Europe were poring over splendid illustrations in books Tycho Brahe had sent them as gifts. No book had arrived for Rudolph, but the reason for this omission was in the air at court: Tycho wished to present it to the emperor in person. Soon Lehmann reported that Rudolph could scarcely contain his eagerness to meet Tycho and would offer him a sumptuous dwelling. In mid-August Tycho heard from his other contacts that the time was ripe for him to come to Prague: The emperor was indeed prepared to extend his patronage.

Even while Tycho's fortunes were improving, Ursus had never been far from his mind. Longomontanus, having just completed his master's degree at Rostock, was back for a time that summer, mainly to discuss the Ursus problem. It gave much cause for celebration when the news arrived from Lehmann that the emperor had lost all confidence in the trustworthiness and abilities of his present imperial mathematician.

By September 29 Tycho, his family, and his retainers had packed everything up for another move, and the great household was ready to depart from Wandsburg. Instruments, library, furniture were on the wagons again, lashed down and protected from the weather. The carriages were brought, and the horses harnessed. Footmen helped Kirsten and the children into their carriages. The armed escorts mounted. Tycho's carriage with its six horses led the way, and the long train of animals, wagons, carriages, and outriders fell into line, rattled back across the drawbridge of Wandsburg, and turned south along the Elbe River, in the direction of Prague.

This time, the journey was more like a royal progress than moving house. There had been time to prepare, have proper clothing made for himself and his wife and children, inform noble and scholarly friends and kinsmen that he would be calling on them along the way. By October 5 the convoy reached Harburg, not far from Hamburg, where Tycho elicited another letter of reference, this one from the aged Duke Otto II of Braunschweig-Lüneburg to the imperial high steward at the court.

When they reached Magdeburg, Tycho took advantage of the presence of an old friend and correspondent, Rector Georg Rollenhagen, to further his campaign to crush Ursus. It was from Rollenhagen that Tycho had first learned about Ursus's earlier book, *Fundaments of Astronomy*, in which Ursus claimed the Tychonic system as his own, and Rollenhagen had put Tycho in touch with Lehmann, his brother, at the imperial court. Tycho now took the opportunity to question Rollenhagen in detail about an incident in 1586 when Ursus, asked to explain his planetary system, had proved mysteriously incapable of doing so—further evidence that he had not invented it himself. Rollenhagen supported Tycho's decision that a nobleman should not sully himself by disputing directly with a pig farmer—no matter how high that pig farmer had risen—but should allow lawyers, clients, and friends to take care of this matter for him.

Tycho had also summoned Erik Lange to Magdeburg. After making it clear he would not cooperate with Lange's "appalling plan" to accompany him to Prague to interest the emperor in underwriting his experiments for turning base metals into gold, Tycho compelled Lange to testify before a notary about Ursus's behavior those many years ago at Uraniborg. By this time Tycho must have all but given up hoping that Kepler would ever respond to his letter of the previous April, which had included a request for a document Tycho could use against Ursus.

The stop in Magdeburg gave Tycho time to consider whether it might not work against his interests to arrive in Prague accompanied by family and twenty-two wagons, as though he were a refugee casting himself on Rudolph's mercy or, on the other hand, an overconfident man taking Rudolph's munificence too much for granted. A lack of total commitment to the move might even help in negotiations having to do with how generous the emperor's patronage would be. If it seemed advisable to retreat for a time, Tycho would have a good excuse, to return to his family and work. Perhaps most compelling of all, the journey would go much more quickly than it could with all the wagons, and winter was not far away. Tycho reverted to an earlier plan of taking a small company of assistants and only a few of his instruments with him. He left the other instruments and much of the baggage train in Magdeburg, and Kirsten and their daughters and servants, escorted by Longomontanus, returned to Wandsburg. Tycho continued the journey with his two sons and a few other retainers, including the superbly effective Tengnagel.

They still did not ride posthaste to Prague, but stayed a month in Dresden while waiting for surer confirmation that a suitable welcome awaited Tycho. Word came that the emperor was delighted to hear he was on his way to Prague, but that an epidemic was raging in the area, winter was closing in, and the court had temporarily moved from the city. Tycho was advised to wait until spring. He decided to winter in Wittenberg, Martin Luther's old city.

15

CONTACT

November 1598–June 1599

IN LATE NOVEMBER or early December 1598, Kepler opened a letter from an extremely agitated Mästlin. This was the first Kepler knew that he had made a fool of himself with potentially disastrous results for his career, and Mästlin's letter did not even make it clear precisely what the situation was, except that it was bad.

The reply Tycho had written Kepler shortly after receiving the two books and Kepler's letter at Wandsburg never reached Kepler. But Tycho had sent Mästlin a copy, and Mästlin, reading that in June and assuming that Kepler had the original, had immediately written to Kepler, reprimanding him for praising Ursus when Mästlin himself had warned him that the man's work was worthless. Then Mästlin's reprimand had also gone astray and taken five months to reach Kepler.

Kepler asked Mästlin to send him a copy of Tycho's April letter, and finally had that in his hands in February. It was not nearly so awful as he had been expecting. It was even rather polite and complimentary. Although Tycho expressed some reservations about the polyhedral theory, he wrote that he found it extremely ingenious and hoped Kepler would try applying it to the Tychonic system. He went on to comment, however, that Copernicus's measurements for the planetary distances were not accurate enough for the purposes to which Kepler was putting them, and Kepler might want to use instead the more accurate observations that he, Tycho, had made.

Tycho's hook dangled enticingly before Kepler's eyes. He was so overjoyed that he penned in the margin, "One can see his high opinion of my method."

There was, however, more to Tycho's letter, a lengthy postscript containing what was, under the circumstances, a remarkably restrained complaint about Kepler's unctuous praise of Ursus. Tycho wrote that he could not imagine that Kepler had been aware that Ursus would use his letter in his "defamatory and criminal publication." He suggested that Kepler might give him a statement of his opinion of Ursus's behavior, which document Tycho could use in legal proceedings against Ursus.

Tycho had given Kepler a glimpse of the reward that might be his and had named the price of his forgiveness. Kepler also knew that in a separate letter to Mästlin, Tycho had criticized *Mysterium* much more severely and complained about the letter reprinted in Ursus's book. Tycho surely anticipated that Kepler's influential mentor would communicate with his errant pupil, and he was treading a delicate line between flattery and encouragement on the one hand and criticism and reprimand on the other.

Even before he had read the copy of Tycho's actual words to him, Kepler had taken Tycho's bait on the strength of Mästlin's correspondence. He had also not seen Ursus's book and did not know whether Ursus had quoted him correctly or distorted what he had written. Kepler could not even remember precisely what he *had* written, nor did he know which letter Ursus had quoted, for there had been more than one, and he had not kept copies. In a blind panic about the appalling situation he had got himself into with one of the great men of the age, who actually was inclined to *like* his work, Kepler threw himself at Tycho's feet in a long reply, written with characteristic eloquence and dramatic flair:

> Why does [Ursus] set such value on my flatteries? . . . If he were a man he would despise them, if he were wise he would not display them in the market place. The nobody, that I was then, searched for a famous man who would praise my new discovery. I begged him for a gift and behold, it was he who extorted a gift from the beggar. . . . My spirit was soaring and melting away with joy over the discovery I had just made. If, in the selfish desire to flatter him I blurted out words which exceeded my opinion of him, this is to be explained by the impulsiveness of youth.

Tycho's response this time was gracious, but he kept Kepler in his place by saying that he had not required such an elaborate apology.

THAT WINTER, Tycho was sumptuously lodged in a famous old house that had once belonged to Melanchthon, with space and time for his studies and chemical work and for attending to his publications. Kirsten and their children joined him. He was surrounded by intellectual friends who set enormous value on his company, and he was temporarily not required to fawn over kings or emperors. Tycho knew that the success of his pilgrimage to seek imperial patronage depended on his continuing to appear frequently and splendidly in public, maintaining a reputation in his own field but also as someone with an extensive knowledge of natural philosophy and the arts. No situation could have suited him better. He was in his element.

Even though Rudolph's patronage seemed almost assured, Tycho had experienced the mercurial nature of kings and knew he could not take it for granted that the emperor would offer to fund him and his research permanently and in as generous a manner as he required. It might still happen that he would be greeted as a distinguished visitor, converse with the emperor, but only be given gold chains and medallions and finally have to return to Wandsburg. Worse yet, the emperor might grant him no audience at all. Power brokerage was such that the word of influential friends and informants at court was almost as good as a contract, but it was nothing signed, sealed, and delivered. Tycho, for all his assurances and letters, would still be approaching Rudolph as a supplicant. Hence Tycho continued to use every means possible to make his position in Prague stronger, not neglecting to curry favor in other courts as well. He dispatched Tengnagel across the Alps to present copies of *Mechanica* and the star catalog to rulers in Venice, Florence, and Parma. Again, Tengnagel did well for himself and for Tycho. The Order of San Marco in Venice gave him a knighthood, and he secured for Tycho the favor of the doge of Venice, Grand Duke Ferdinand de Medici in Florence, and the Farnese family of Parma, all of whom had close ties with the court in Prague.

There had been a belated flicker of hope for Tycho in Denmark the previous autumn. King Christian had married Princess Anne Catherine of Brandenburg, and Viceroy Rantzau had attended the wedding. On their journey home in mid-December, the bride's parents, Margrave Joachim Frederick and Margravine Catherine of Brandenburg-Küstrin, passed through Schleswig-Holstein, and Rantzau arranged for Tycho to meet them. So convincingly did Tycho and Rantzau make their case that the new royal in-laws agreed to take up Tycho's cause with their son-in-law Christian. In January, the margrave became elector of Brandenburg, which meant he was even better positioned to exert influence. He and his wife wrote to Christian. They even took care to disguise their letters as personal family correspondence so that they would not be intercepted, for Tycho knew he had enemies at the Danish court and even suspected that the reply to his earlier appeal had not been written by King Christian himself. The margrave's and margravine's letters nevertheless fell into the wrong hands and were too much delayed to be of any use.

By this time Tycho, with his eyes set on Prague, was less discouraged by this setback than by another having to do with his astronomy. He heard from Longomontanus, who had returned to his boyhood home in Jutland, that he had observed a lunar eclipse from there in late January, and it had occurred half an hour earlier than Tycho had predicted. Tycho's plan to present a book about his lunar theory to the emperor when he had his first audience had to be scrapped in favor of another manuscript about daily solar and lunar positions for 1599. Nevertheless, Tycho's hopes were finally reaching fruition. By March he had his summons: His Imperial Majesty desired to receive him in Prague.

Tycho had bookbinders create a presentation copy of the book of solar and lunar positions. He wrote to Longomontanus, then in Rostock, asking him to come quickly and travel with him. He bought new, superb carriage horses. In his chemical laboratory he cooked up medicines against plague and other diseases to take with him as gifts and for his own use. The journey to Prague was delayed again when Magdalene became seriously ill, and then delayed yet again because the spring thaw and heavy rains made the roads impassable.

At last, on June 14, 1599, Tycho and his entourage left Wittenberg, overnighted with lavish entertainment at Castle Pretzsch along the way, and arrived at Dresden, where Kirsten, their children, and the servants would wait in the household of a high Saxon official. Tycho, in his finest carriage with the new horses, accompanied only by his eldest son and servants—for Longomontanus had not yet arrived—proceeded to Prague. Early in July, as they approached the city, Ursus slipped away.

❧

IN GRAZ, Kepler heard about Tycho's triumphal entry into Prague. He also heard about Ursus's ignominious departure. He had no idea whether these events had any relevance to his own prospects.

The only thing that could be said was that Tycho had come nearer. It was no longer completely out of the question that Kepler might someday travel to meet him. The successful continuation of Kepler's work seemed to him increasingly to depend on being able to consult observations that only Tycho could supply.

The summer of 1599 was a tragic one for the Keplers. Their second child, Susanna, born in June, lived only thirty-five days. Kepler refused a Catholic burial for his infant daughter and was fined for his stubbornness. When he appealed, the fine was lessened, but he still had to pay it and was forbidden to bury the tiny body until he did.

As he grieved for his child, tried to comfort his wife, and worried about how long they could hold on in Graz, Kepler's mind was also busy in a new and happier direction. He wrote to Edmund Bruce, an English acquaintance living in Padua, about a new theory he thought might interest Bruce's friend Galileo, and he described the theory in letters to Mästlin and Herwart von Hohenburg. In his search for answers to the questions that interested him, Kepler had begun to look to music. He was not being illogical, mystical, or, for that matter, original.

One of the most profound and far-reaching intellectual advances in the history of human knowledge occurred in the sixth century B.C. when the Pythagoreans recognized that there are fundamental mathematical relationships hidden in nature. It was in music that they made this discovery, and it seemed to them, with ample reason, a glimpse into the mind of God. Plato also recognized that musical harmony is a manifestation of deep mathematical relationships, and he, as Kepler would later, speculated about the possibility that the arrangement of the cosmos might be another of those manifestations.

Tycho had been thinking along the same lines when he designed Uraniborg: Plucking the string of a harp produces a musical tone—the "ground note" or "fundamental." If one presses the string down at its center—creating, in effect, two strings each half as long as the original one—and plucks again, the tone is an octave higher than the fundamental. A ratio of string length one to two produces the octave; two to three produces a fifth; three to four, a fourth; four to five, a major third; five to six, a minor third; three to five, a major sixth; and five to eight, a minor sixth. The ancients preferred only the ratios 1:2:3:4, but Kepler learned of more modern music that used the other ratios as well. All these ratios produce musical intervals that human ears find "harmonious." There is some deep connection between these lengths of string, these ratios, and the human mind.

For Kepler this was not at root a matter of mathematics or numbers. He was no numerologist. For him, the most fundamental attribute of nature was geometry, a clearer and simpler demonstration than mathematics that some things are possible and others are not—that some things "fit," and others do not. It was the link between geometry and musical harmony that intrigued him.

Kepler's polyhedral theory failed to explain two things about the solar system about which Kepler was particularly curious. The first was the size of the planetary "eccentricities." In Ptolemy, when a planet's orbit was not precisely centered on Earth, but rather on a point near Earth, the orbit was said to be "eccentric." Similarly, a Copernican like Kepler called a planetary orbit eccentric if it seemed to be centered not precisely on the Sun but near it. Kepler wanted to know how eccentric the orbits of the planets were, but he also wanted to know the physical reasons why a planet had a certain degree of eccentricity and not another.

The second question his polyhedral theory could not answer was how a planet's distance from the Sun was related to the length of time it took to complete one orbit, known as its period. Kepler had discussed that relationship in *Mysterium,* but not to his satisfaction. Again, he wanted to know not just *how* they were related but also the physical causes *why* this must be so. Kepler continued to adhere to his belief that God's creativity lay deeper than merely setting things up and moving them around. God, he was certain, had established an underlying logic and perfect harmony from which all things had to proceed. Kepler had already found that discovering pieces of this logic was pure delight, which he felt must echo God's delight in creating. Since musical harmony was a manifestation of that same divine logic, it seemed a highly promising area in which to find links with the cosmos.

His first results seemed to indicate that he was right. In his letters to Bruce, and to Mästlin in July 1599, and a little later to Herwart von Hohenburg, Kepler pointed out that his new "harmonic theory" gave more accurate predictions than a theory he had earlier included in *Mysterium*. He was suggesting that the planets, moving through something similar to air, made sounds the way the strings of a lyre do if the lyre is hung up so that the breeze moves across them. The velocities (the "vigor," as he put it) of the six planets might be related to each other in the same relationships that would produce a harmonious chord if one translated them into lengths of strings on a stringed instrument. For example, say the relationship between the velocities of Saturn and Jupiter was 3:4. On a stringed instrument a 3:4 ratio between string lengths produced the musical interval of a fourth. It followed, said Kepler, that the "interval" between Saturn and Jupiter could be thought of as a musical fourth. Kepler worked out this scheme with all the planets' velocities. He calculated the proportions of their velocities as follows: Saturn to Jupiter, 3:4—a fourth; Jupiter to Mars, 4:8 (or 1:2—an octave); Mars to Earth, 8:10 (4:5—a major third); Earth to Venus, 10:12 (5:6—a minor third); Venus to Mercury, 12:16 (3:4—a fourth). From all these intervals, Kepler built up a chord, a C major chord in what musicians call its "second inversion (see figure 15.1)." Kepler was not entirely pleased with this second inversion. He would have preferred the chord to be in root position—a C major chord with C as the lowest note. However, he wrote with a shrug, "thus it is in the heavens."

Having chosen the velocities with the goal of creating a harmonious chord, Kepler was encouraged when he found that the *musical* intervals were not far off from the *spatial* intervals between the plan-

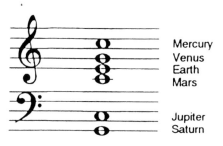

Figure 15.1: Kepler's 1599 planetary chord.

ets in his polyhedral theory. The planetary periods—the amount of time it took each planet to complete its orbit—were well known. He set out to calculate how large the different planetary orbits had to be in relation to one another if the planets, with these periods, were traveling at the velocities his musical intervals predicted. Then he compared his results with the orbital sizes calculated from Copernican principles. Though still far from perfect agreement, his harmonic theory was in somewhat better agreement than his polyhedral theory.

Kepler proceeded to sort out what he had learned from each theory. The harmonic theory allowed him to figure out the planets' distances from the Sun, relative to one another. The polyhedral theory gave him the thickness of the empty spaces between the spheres in which the planets moved. He thought that the space left over might be the space required by the planetary eccentricities, for a sphere had to be thick enough to accommodate the eccentricity of the planet's orbit (see figure 15.2). The task then was to figure out how to distribute that leftover space among the different spheres. It seemed possible that the answer to the question of why the eccentricities of the planets' orbits were what they were was that any other eccentricities would spoil the harmony.

Kepler soon wrote to Mästlin about a clever way he had found to relate his arrangement of the Platonic solids to three of the five intervals

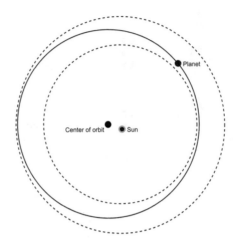

Figure 15.2: In this drawing, the borders of the planet's sphere are shown by broken lines, the orbit by a continuous line. The sphere of a planet had to be roomy enough to accommodate the planet's eccentricity. In other words, the planet had to be able to travel in the parts of the orbit where it was nearest to and farthest from the Sun without breaking out beyond the borders of its sphere. The more eccentric the orbit, the thicker the sphere needed to be.

in his chord. It was the sort of connection Kepler loved. The polyhedron that separated the spheres of Saturn and Jupiter in his theory was a cube. Each corner of a cube is the meeting place of three flat squares, and the corner of each of the three squares is a ninety-degree angle. Add those three angles together, and the sum is 270 degrees. The ratio between 270 and 360 (the number of degrees in a complete circle) is 3:4. Thus it seemed appropriate to Kepler that the musical interval that requires a ratio of string lengths 3:4 should define the space interval between Saturn and Jupiter. Similar relationships worked for the interval between Jupiter and Mars and between Earth and Venus.

Kepler confided to Mästlin and Herwart von Hohenburg, later in the summer, that having come thus far with his harmonic theory, he felt as though he had "a bird under a bucket." His sense was that the harmony he was suggesting reflected the mind of the Creator and therefore surely had to be carried out in the cosmos. Kepler thought he was close to an answer. In fact, it would be twenty years before he found it, and when he did, he would also have discovered his third law of planetary motion.

In the meantime, Kepler's great frustration was that being so near, he believed, to resolving the discrepancies in his theories, he lacked precise, accurate values for the observed eccentricities of the planets and the dimensions of the solar system. Kepler had no astronomical instruments of sufficient quality to make the necessary observations himself. The only man in the world who did was Tycho.

Kepler also longed to know whether Tycho's observations revealed any tiny differences in the altitudes of the Pole Star at the winter solstice and the equinoxes. The absence of any stellar parallax shift was still a difficulty for a Copernican, because the shift ought to be visible as Earth orbited the Sun. Kepler remained troubled by the fact that Copernican astronomy, combined with the absence of stellar parallax, required the stars to be so inconceivably far away.

At the same time, Kepler had little respect for the "patchwork" Tychonic system, no matter who had thought of it, Tycho or Ursus. To Mästlin he wrote of Tycho's ideas about the arrangement of the planetary system as "little paper houses" and wondered whether his own theory of celestial harmony might not easily blow them over. However, Kepler also wrote that he was reluctant to contradict a man like Tycho, especially since he needed Tycho's observations. If only, he complained, Tycho would *publish* them so they would be widely available, or, failing that, be persuaded to send Kepler the information he needed. "I did not wish to be discouraged, but to be taught," Kepler complained to Mästlin, rankled by the reservations about *Mysterium* that Tycho had expressed. Kepler could hardly contain his impatience. "My opinion about Tycho is this: he has abundant wealth. Only, like most rich men, he does not know how to make proper use of his riches. Therefore, one must take pains to wring his

treasures from him, to get from him, by begging, the decision to publish all his observations without reservation."

To Kepler, Tycho was coming to resemble the dragon nesting on a
hoard of gold, not able to put it to use in a meaningful way himself,
not even recognizing its true value, but too fearful of thieves to allow
anyone else to glimpse it. Kepler was not above putting his mind to
scheming. There must be other ways to get at this hoard than "by
begging the decision to publish."

16

PRAGUE OPENS HER ARMS

July 1599 – February 1600

TYCHO BRAHE'S RECEPTION in Prague quickly allayed all fears that his journey might have been in vain. Johannes Barvitius, the emperor's private secretary and one of the triumvirate who had brokered Tycho's relationship with the imperial court, met him in a garden near Rudolph's palace. Tycho showed him the three books he had brought for Rudolph and the letters of introduction. Barvitius said he would find out in what way and from whom Rudolph wished to receive them. The answer Tycho hoped for came a day later: The emperor wanted to receive them from Tycho himself, and he would be summoned into Rudolph's presence shortly.

Meanwhile other high-ranking officials at the imperial court welcomed Tycho warmly and expressed their outrage that King Christian had so sadly undervalued his achievements. Tycho replied diplomatically by defending the Danish king and praising his talents. When the discussion turned to the ignorance and villainy of others at the Danish court, Tycho was only slightly more willing to agree but quickly turned the conversation by commenting that "perhaps God has acted by some special providence in order that the astronomical investigations with which I have been so long and so thoroughly occupied should now come elsewhere and redound to the credit of the emperor himself."

Barvitius drove Tycho in his carriage to "a splendid and magnificent palace in the Italian style, with beautiful private grounds," situated on the pinnacle of the same hill on which Rudolph's glorious, sprawling complex stood. Barvitius pointed out the advantages of the mansion, including a tower that might serve for astronomy, and told Tycho that if he liked it, the emperor was willing to purchase it for him.

While they toured the house and grounds, Tycho found subtle ways of letting Barvitius know "from what I said and did not say" that the tower was inadequate to hold even one of his instruments and that he was not overjoyed with the house. The emperor had foreseen this possibility, and Barvitius was immediately able to mention several castles outside Prague, reachable within a day or two, where there would be fewer disruptions, fewer envious eyes, and a situation nearer to what Tycho had enjoyed at Uraniborg. Barvitius also informed Tycho that Rudolph was prepared to give him an annual stipend, which Tycho would hear more about when he had his audience.

When the summons came, Tycho had the rare honor of entering Rudolph's audience chamber alone. "I saw [the emperor] sitting in the room on a bench with his back against a table, completely alone in the whole chamber without even an attending page. After the customary gestures of civility, he immediately called me over to him with a nod, and when I approached, graciously held out his hand to me. I then drew back a bit and gave a little speech in Latin." Rudolph replied with equal grace, "saying, among other things, how agreeable my arrival was and that he promised to support me and my research, all the while smiling in the most kindly way so that his whole face beamed with benevolence. I could not take in everything he said because he by nature speaks very softly." Tycho thanked the emperor and excused himself to fetch the three presentation copies of his books that he had left with his son Tycho in the antechamber.

Rudolph "took them and laid them out on the table. I reviewed the contents of each briefly. Then he again responded with a splendid speech, saying most graciously that they would please him greatly. I then removed myself according to the proper courtesies."

Rudolph called Barvitius into the audience chamber. Barvitius emerged again almost immediately to tell Tycho that Rudolph had been watching from his window as Tycho arrived and had noticed a mechanical device on Tycho's carriage. He now wished to have it shown to him. The device was Tycho's odometer. Tycho ordered his son to fetch it and gave it to Barvitius with a quick explanation of its construction and operation. Barvitius soon came back out of the audience chamber to report that the emperor did not want to accept Tycho's odometer but would have one made for himself according to its pattern. This was Tycho's first experience of one of the emperor's eccentricities. Rudolph was a fanatical collector of curiosities and was far happier among these objects than among people.

Barvitius reported that "the emperor was very favorably disposed toward [Tycho] and that after he referred the case to his council, in a short time, he would settle the matter of an annual grant and suitable quarters." Tycho was encouraged to summon his family to join him, and the "emperor himself would do everything necessary to make sure that we lacked for nothing needed to live comfortably." Tycho sent his son to bring Kirsten and the others from Dresden. They all arrived eight days later.

It was urgent that Tycho have an estate that he could begin turning into a new Uraniborg, and, true to his word, Rudolph gave Tycho a choice among three estates some distance from the city, including his own favorite hunting lodge, Brandeis—a huge, magnificent establishment. Perhaps thinking of what a sacrifice the emperor would be making if he accepted it, Tycho chose another castle farther along the same road, a six-hour ride from Prague. Handsomely positioned on a bluff sixty meters above the flood plain of the river Jizerou, this was the castle Benatky, the Czech name for Venice, because when the area flooded, the promontory where the castle stood was surrounded by water. The mansion, like Uraniborg, was not ancient and had never been intended to serve as a fortress. Also like Uraniborg, it boasted an indoor water system, probably the first in Bohemia. Tycho was taken by the beauty of the surroundings and by Benatky's uninterrupted view of the horizon in all directions, and he even noted with approval that there was, nearby, a small village of Protestants with Calvinist leanings.

By late August Tycho and his family had explored the bright, spacious rooms of the castle and were deciding how they should be allotted and where the furniture should go. The indoor space was larger than Uraniborg, with three floors of nearly 5,380 square feet each. Though none of this space was well suited for astronomy, Tycho was setting up his instruments. Rudolph had by now had time to peruse the pictures in Tycho's *Mechanica,* and he was eager to see Benatky become even greater than Uraniborg.

Tycho began the transformation of the castle by making repairs, but it was not long before he was modifying the floor plan and windows and designing additional buildings to house the instruments and alchemical laboratories. He took observations to calculate the exact geographical position and orientation of his new home, and he marked the meridian with a line on the floor near a window. Benatky wasn't oriented along north-south, east-west lines, as Uraniborg had been, but then it had not been designed by an astronomer.

Tycho's annual grant took longer than expected to pass the council. The outlay for Tycho was to be higher than the salaries of many counts and barons in the emperor's service. Eventually Rudolph even ordered that Tycho's salary be retroactive from the time his patronage in Denmark had ended, and Tycho was to have a hereditary fief as soon as one became available. However, shortly after Tycho began his remodeling, the administrator of the estate, Caspar von Mühlstein, began complaining to Barvitius about the mounting costs. By late November, Tycho's renovation estimates had doubled, and

Mühlstein had also learned that the salary the emperor had promised Tycho was much greater than the income from the Benatky estate. Mühlstein refused to authorize any more expenditure without an official order backed up with money from the treasury. He knew, as Tycho would soon discover for himself, that much of Rudolph's munificence was, in fact, financial make-believe. Unlike in Denmark, where the king's word was his bond and bound everyone else as well, in the imperial court promises often rested on nothing but good intentions, orders on the royal treasury would fail to produce payment, and there might not even be sufficient money in the treasury to make good on Rudolph's pledges.

Tycho soon found that in other respects as well the emperor's favor did not make all things possible. The promised hereditary fief could not be Tycho's until he had applied for and obtained citizenship, a slow bureaucratic process. His friends warned him that envy and slander were as much a part of court life here as they had been in Denmark, and there were opportunists eager to bring him down. Powerful men who had not been part of Tycho's network were not pleased to be shouldered aside by a foreigner.

However, it also was not long before officials and administrators such as Mühlstein began to realize the seriousness of Rudolph's intentions to underwrite Tycho's work. The court had again left Prague for fear of plague, taking with them some of Tycho's medicines, but letters passed frequently between Tycho and the emperor. Tycho's messages went straight to Rudolph without perusal by the Imperial Council. With the emperor giving him this much priority, Tycho felt so confident that he threatened the foot-dragging Mühlstein with Rudolph's displeasure and hinted that if the expenditures were not authorized he might "leave Bohemia and tell the world why." The Chamber of Deputies informed Mühlstein that, awaiting Rudolph's clarification, he should continue construction at Benatky as cheaply as possible.

Rudolph's response on December 10 set matters straight: Tycho was to have his wooden outbuildings and "little rooms," bays along the cliff for the instruments. Tycho's salary also was to be paid, in part out of rents from Brandeis.

By late autumn so much remodeling was going on at Tycho's castle that there was less living and working space than there had been to start. Also, the plague had come nearer, and two thousand had died in the district. Tycho moved his family to another castle twenty miles downriver because "the women were frightened," as he reported it.

All this activity had not made Tycho forget about Ursus. In September he had begun investigating the man's whereabouts and learned of his flight from Prague. Tycho secretly consulted the official censor. Because Ursus had published his book without first submitting it to censorship, it was within the censor's power to summon him and assign punishment. But with the imperial court absent from the city, there was no court before which Ursus could be summoned. Tycho had to wait, but he reaffirmed his intention to track "the beast" down and drag him out of hiding.

❦

THAT AUTUMN, Johannes Kepler and his family faced an increasingly ominous situation in Graz. Kepler could no longer escape into mathematical and philosophical speculation and ignore the threat hanging over him. There were rumors that soon any Lutheran moving away from Graz might not be allowed to take away his possessions or trade or sell them, confirming the fear that had earlier made the Keplers decide to try to weather the storm rather than relocate elsewhere. The loss of Barbara's substantial inheritance and all her possessions would have been catastrophic.

Nevertheless, to stay was becoming untenable. Oppressive ordinances touched the family directly, and forced conversions to Catholicism were surely not far away. Riots broke out continually in the city and nearby countryside. "No matter what fate might await me if I move elsewhere," wrote Kepler, "I know for certain that it will not be worse than that which threatens us here so long as the present government continues."

He did not have many options. Returning to Württemberg to take up a clerical position, the ambition he had painfully relinquished when posted to Graz, was out of the question, because his disagreement with Tübingen orthodoxy, begun as a student, was now stronger than ever. "I could never torture myself with greater unrest and anxiety than if I now, in my present state of conscience, should be enclosed in that sphere of activity," he wrote.

One possibility was a university professorship in philosophy or even in medicine. Kepler appealed to Mästlin, asking whether there might be a position like that available at Tübingen or whether he should look elsewhere. He inquired about the cost of living in Tübingen—the price of bread, wine, and rent. Mästlin replied that, sadly, he had no advice to offer and lamented that Kepler had not sought the counsel of a wiser man with more political experience, "for in these matters I am as innocent as a child." He reported the prices of grain and wine but advised Kepler not to hope for a future in Tübingen. Kepler also wrote to Herwart von Hohenburg, who had been such a helpful friend in the past. Von Hohenburg failed Kepler this time. His own position was insecure, and he needed to be exceedingly discreet.

With each failure, Kepler's thoughts returned to Tycho Brahe, whose success in Prague had been reported to him by von Hohenburg. Prague was not, after all, so far away. Tycho had mentioned the possibility that Kepler might like to use his observations. The idea that that suggestion could possibly be construed as an invitation or even as a job offer seemed tempting but outrageous. Kepler would have to leave his family behind in Graz, for the letter had said nothing of them, and he would have to put his own mind and talents at the disposal of another man, rumored by some to be a tyrant. He had already offended that man, and Tycho's forgiveness had been gracious but condescending.

With the new year, 1600, and a new century, came the invitation from Johann Friedrich Hoffmann, baron of Grünbüchel and Strechau, member of the diet of Styria and councillor to Emperor Rudolph. Hoffmann offered Kepler not only a way to get from Graz to Prague, but an introduction to Tycho Brahe. Kepler's wavering ended.

Kepler would have begun the journey in Hoffmann's carriage with much more confidence had he known about a letter from Tycho that arrived shortly after his departure. Tycho had repeated his invitation—and this time it clearly *was* an invitation, insisting that Kepler must come to Prague, not because he was "being forced out of Graz" but of his own free will and because he "desired joint studies" with Tycho. If he chose to come, Tycho was prepared to help and advise him and his family.

Kepler and Hoffmann's ten-day journey from Graz to Prague passed through rolling countryside studded with promontories, many of which were crowned by castles. Prague itself was situated around such a promontory, with an enormous castle complex that even included a cathedral. This complex, a great city within a city, was the seat of the Holy Roman Emperor. Renaissance mansions of some of Europe's most powerful aristocrats lined the higher parts of the steep streets that climbed the hill from the river called the Vltava by the local population and the Moldau or Moldova by their rulers. Farther down, nearer the river, were the homes of courtiers and craftsmen, and there was much more of the city at the other end of a long, stone, towered bridge that spanned the river.

Kepler, still unaware of Tycho's second invitation, stayed for a few days as guest of Baron Hoffmann. Cosmopolitan Prague was a different world from the Graz Kepler had left. Bustling, exhilarating, a mixture of narrow, malodorous streets, wider avenues, and broad marketplaces, it was alive with many ethnic groups and languages. The court attracted a diverse community of noblemen and ambassadors from all over Europe, as well as opportunists and hangers-on, and this community in turn provided a living for hundreds of trades-men, craftsmen, scholars, and artists.

It was some days before Kepler was able to get word to Tycho that he was in the city. However, "as soon as I arrived," Kepler reported, he had an unpleasant encounter with Ursus, who had not fled far after all. At first Kepler kept his identity a secret from Ursus, "lest he intensify the situation to a brawl," but he spoke to the older man sharply about how little he liked Ursus's recent book. As the incident continued, Kepler let Ursus know who he was and told him that "since he decided to drag me, who had written as a pupil, unwillingly into the judge's chair, he should therefore permit me to discard a pupil's modesty and assume a judge's authority in this literary contest and in my turn de-cide publicly what seems to be the mathematical issue."

By January 26 Tycho, back from his brief flight from the plague, had heard of Kepler's arrival. He wrote to Kepler again with extreme cordiality: "You will come not so much as guest but as very welcome friend and highly desirable participant and companion in our obser-vations of the heavens." He sent his son Tycho and Tengnagel in his own carriage to Prague with instructions to bring Johannes Kepler back with them. Kepler had every reason to anticipate from Tycho as warm and accommodating a welcome as Tycho had received six months earlier from Emperor Rudolph.

17

A DYSFUNCTIONAL
COLLABORATION

1600

THE CARRIAGE in which Kepler rode must have creaked, shifted on its axles, and tilted as the horses began the steep pull up the road to the top of the bluff where Benatky Castle stood. Kepler's mood cannot have been other than one of excitement and exhilaration, with most qualms about the impending meeting overridden by anticipation that Tycho Brahe, better than any other man alive, would be able to understand and value his ideas. The intellectual relationship and the access, at last, to Tycho's observational data that Kepler looked forward to, and that Tycho had promised in his letters, must surely have made the future appear as rich, fertile, and limitless as the plains and skies that opened to view as the road climbed.

Tycho's arrangement for Kepler to ride from Prague in the carriage with his eldest son had been flattering and consistent with the tone of his most recent correspondence, and Kepler's welcome at Benatky was no disappointment. The venerable astronomer granted him a cordial initial interview. Tycho Brahe's mystique was as powerful as any monarch's, and to Kepler he probably seemed like a character from legend who had turned out to be real. Kepler reported that Tycho offered to reimburse his travel expenses, and he "saw immediately" that there was "no fear that I would regret the trip."

Alas, it was not long before Kepler's mood deteriorated to bleak disillusionment, homesickness, and panic about the future. The promise of that welcome turned out to have been a cruel mirage. In the days that followed his arrival, Benatky's harried lord turned to other matters. The bustle and confusion of a castle under reconstruction went on around a bewildered Kepler as though he were not there. Perhaps he should not have been surprised. Only recently he had visited the court in Württemberg and barely been allowed to sit at the Trippeltisch. Tycho's households at Uraniborg and Benatky, though they in some ways resembled the establishment of a university professor, still had much in common with the court of a feudal ruler, where a man of lower status rarely had contact with that ruler except for the sight of him at the dining table.

Kepler's plight was not helped by the fact that Tycho was in an unusually distracted state. He was worried about his instruments, not only the four left on Hven but the others that were still in Magdeburg. He was trying to supervise the renovation of a castle teeming with workmen who needed his direction, for in his mind alone existed the vision of what the reconstruction would look like when it was completed. He was coming to grips with the huge disappointment of learning that a promise of financial support from the emperor did not mean that money would be forthcoming. Not for Tycho, nor for Kepler—nor for anyone else at Benatky—were circumstances conducive to systematic and productive scientific work. Tycho, like Kepler, was homesick, frustrated, uncertain of the future, and on the verge of financial disaster.

At first it seemed there would be nothing for Kepler to do that he personally could consider worthwhile. There were workmen to be dodged morning, noon, and night; there were women—Tycho's wife and daughters and their attendants—chattering in Danish (which Kepler, of course, did not speak) as they, too, tried to live normally amid uprootedness and turmoil. In spite of the crowded conditions, Kepler called the situation "a reigning loneliness of people." He particularly despised mealtimes—boisterous, rowdy, claustrophobic for a man accustomed to the peace and quiet of his own home. The meals were served in a room on the second floor of the castle, with Kirsten and their family joining Tycho and whatever assistants and visitors were currently in residence. Kepler had never, even in his student days, drunk as much wine as they did at this table. Yet these uncomfortable mealtime intervals were his only opportunity to learn anything at all about the observations he desperately wanted to consult. Sometimes, as casually as he might cast a bone to his dogs or a tidbit of sweet to his jester, Tycho would come forth with a snippet of precious astronomical information; "One day . . . the apogee of one planet, the next day the nodes of another!" grumbled Kepler. And as for the collegiality Tycho had promised him—the discussion of "lofty topics, face to face in an agreeable and pleasant manner"— it seemed Kepler could do no better than occasionally command Tycho's attention between bites for a few fleeting words.

Perhaps Kepler should have paid more heed to the way Tycho had referred, in his letters, to his most highborn assistant, Tengnagel. He had called him *domesticus*—servant or employee. A "hired hand" was the way Kepler later described his own status at Benatky in a letter to Mästlin—and a rather junior hired hand at that. Kepler soon realized that though he had been led to believe he would be treated as Tycho's esteemed colleague, he needed first to climb his way up the ladder in competition with other "assistants." His competitors included some with considerable skill and seniority, like Longomontanus and Tengnagel, as well as Tycho's own son. Kepler was the intellectual match of any of them, yet a month after his arrival he had not improved his position. There was even more competition when Johannes Müller, mathematician to the elector of Brandenburg, arrived, bringing his family with him. It seemed that Tycho had brought Kepler here only to waste his time: "One of the most important reasons for my visit to Tycho," Kepler wrote to von Hohenburg, "was the desire to learn from him more correct figures for the eccentricities in order to examine my *Mysterium* and *Harmony*. But Tycho did not give me the chance to share his knowledge." Kepler could not at first understand why Tycho was being so secretive.

Though Kepler may have felt invisible and slighted, Tycho had not forgotten this young scholar who looked so ill at ease at his dinner table. Kepler did not realize, may never have realized, the lingering consequences of the foolish letter he had written Ursus. Kepler's future probably hung at the moment less on his mathematical skills and possible scholarly value than on Tycho's hope of using him as a weapon against Ursus, and his lingering unease about the possibility that Kepler might secretly be Ursus's ally. Though many of Tycho's acquaintances had advised him that he was making too much of the Ursus affair, Tycho could not let it go. That was both fortunate and unfortunate for Kepler. It meant Tycho had to go out of his way to keep Kepler at Benatky, under his own watchful eye, but it also meant that Tycho was extremely reluctant to allow Kepler to see the observations.

Kepler had not yet reached his full potential as a mathematician or interpreter of observations, but that potential had been sufficiently evident from *Mysterium* (Tycho had called the polyhedral theory "quite a brilliant speculation") for Tycho to realize, even before he met Kepler in person, that, properly harnessed with regard to wilder flights of fancy, Kepler might be the assistant who could interpret the Mars observations and discover ways to show that the Tychonic system was correct. Though his Mars parallax campaign had so far been a failure, Tycho's letters from Benatky in 1600 and 1601 show that he had not given up hoping that he—rather than Copernicus and certainly rather than Ursus—might be celebrated for generations to come as the man who provided the true breakaway from Ptolemaic astronomy and set the course for the future. It had to have seemed the worst luck and damnable irony that Kepler, who might have served him so well in this effort, should at the same time represent a double threat. Allowing him to see the precious observations might be tantamount to giving them away to Ursus. Or, just as threatening, it might end in confirmation of Copernicus's system rather than Tycho's. However, paranoia or no paranoia—and in spite of his knowledge that Kepler, at present, preferred Copernicus—Tycho was short of assistants and desperately needed someone with Kepler's gifts. "An astronomical treasure accumulated by the expenditure of sweat and money over so many ears," Tycho lamented, was going to waste.

Kepler himself clearly recognized Tycho's great need for him, which made the older man's secretiveness all the more puzzling. "Tycho has the best observations—that is to say the materials to put up the building—and he has the workers. He is lacking only one thing, the master builder who can make use of all that. Although he is clearly of an extremely architectonic mind, the variety and the depths of the truths with which he is dealing are hindering Tycho, because he is approaching the age of an old man. His mind and all his powers are weakened and are going to be weakened more in a few years so that he will hardly be able to carry out everything alone."

No doubt with terrible misgivings, the old dragon decided to loosen his grip on his hoard just a bit, and Kepler found himself with an assignment. Tycho put him to work—"by divine providence," as Kepler later described it—on the Mars observations under the supervision of Longomontanus, who had rejoined Tycho only about two and a half weeks before Kepler came. Kepler reported that Tycho "saw that I possess a daring mind [and] thought the best way to deal with me would be to give me my head, to let me concentrate on the observations of one single planet, Mars." He would of course be working only as assistant to an assistant, but Longomontanus was almost certainly the most gifted astronomer at Benatky besides Tycho himself. He was a warmhearted, well-meaning, unpretentious man, a

An engraving of Christian Longomontanus by Simon de Pas, 1644.

favorite with Tycho and his family. Furthermore, Tycho usually had his assistants work in pairs. When Kepler asked Tycho whether he could use the true Sun, rather than the center of Earth's orbit, as the reference point for Mars's orbit, and Tycho agreed, Kepler found himself floundering with the mathematics. It was Longomontanus who pointed out a clever technique that Tycho had outlined in a private handbook for his assistants.

Kepler still felt frustrated. He had thought he would be working on his own theories here, not just Tycho's. In order to move ahead with them, he needed time off from Tycho's projects, and he needed data on *all* the planets, not just Mars. "I thought I would later also get the other observations," he said, but that did not happen. He was not finding answers at Benatky to the questions that most intrigued him.

Tycho soon recognized that it made no sense to have his two most able assistants both working on the Mars observations. Kepler was more than capable of carrying on the work on his own. By now Tycho had had a little experience with Kepler in person, which must have confirmed both the good and the bad (as Tycho perceived them) that

Tycho had discerned when he read *Mysterium*. Kepler was at least as gifted as Longomontanus and a better writer. (Tycho praised his "well-rounded way of expression.") Furthermore, Tycho knew of no other assistant available who was better equipped to analyze his observations and find out whether they validated the Tychonic system. "Tycho was pleased with this work I did," wrote Kepler. "He said that he himself had been occupied with similar [work], but liked to avoid the complicated calculations and learn of other people's views." However, by allowing Kepler to be his "master builder," taking the road to discovering whether he was right, Tycho would also be opening the possibility of learning he was wrong, and having the world learn it. His most valuable trump card was that he *alone* had observations—and the means to make more—that could prove or disprove *any* current theory.* This was an advantage he would forfeit if he gave someone as skilled and perceptive, and independent, as Kepler the key to the treasure house. But it was a dubious advantage to a man who had spent a lifetime looking for real answers.

Tycho was not yet prepared to give Kepler full access to his observations, but it was nevertheless a significant step when he reassigned Longomontanus to a project Longomontanus had begun earlier, the lunar theory, and allowed Kepler to analyze the Mars observations without further supervision. Tycho could do little to avert the danger that the young man, intentionally or, worse yet, honestly, might analyze the data and find it supporting Copernicus, not Tycho. But Tycho was running out of options. He did take a rather lame precaution with regard to the threat of Ursus. He extracted a written pledge from Kepler that he would never reveal any of Tycho's secrets.

The tragedy of Tycho's paranoia was that there was definitely at least a part of him that liked Kepler, appreciated his talent as perhaps no other but Mästlin could, and longed to have him as friend and col-

---

*Gingerich and Voelkel have pointed out that Tycho, had he kept his observations strictly to himself, could have claimed whatever he pleased about his findings, whether true or not, and no other scholar had data to gainsay him.

league. Had there been no Ursus, and had Tycho been more open-minded about the Copernican system, the time they spent together might have been a supremely happy and productive time for them both. Instead, Kepler became increasingly out of sorts and unraveled that spring. He had arrived at Benatky with confidence in his own abilities and in the importance of his ideas. Tycho obviously had the intellectual capacity to value these, and also the wherewithal to save the Keplers from their intolerable position in Graz. Yet here was no sympathetic mentor, no secure and rewarding job, only an aloof, suspicious curmudgeon and a hand-to-mouth existence. Kepler had always been vulnerable to the opinion of others. When he was younger, he had written about himself, "My greatest worries are not how to live, not my external appearance, not hurt, not joy, not work itself, but the opinion of other people of me, which I wish only to be good. Whence comes this foolishness about 'being seen'?" How distressing must have been the image of himself he saw reflected in Tycho's eyes, and the confusing way it shifted between liking and suspicion.

As the days passed, Kepler found there was, after all, some progress he could make at Benatky on his own theories. The Mars observations alone *were* sufficient for work on his hypothesis that a force from the Sun moves the planets.

He was looking for some hint in the geometry of Mars's movements that it was responding to a force coming from the Sun, and encouragement was not long in coming. He found, as he had suspected, that the only way he could make sense of Mars's orbit, as it was showing up in Tycho's observations, was to take into account the actual position of the Sun rather than using the center of Earth's orbit as his reference point. This use of the "true" Sun made sense if the Sun was the source of Mars's motion.

Next Kepler used the Mars observations in an innovative way to study Earth's orbit. In writing *Mysterium,* when he had been using older observations, Kepler had been forced to admit, reluctantly, that his idea that the Sun was the source of the movements of the planets seemed not to work for Earth. Using Tycho's observations, he found that it *did* work. Earth like the other planets sped up as it came nearer the Sun and slowed down as it moved farther away. This was a particularly significant and welcome breakthrough.

When Tycho learned of Kepler's findings, he was not happy about them, and he raised objections. Here, as he had feared, was the Copernican system rearing its ugly head. Furthermore, he, like Mästlin, disapproved of Kepler's insistence that in order to develop planetary theory one must find the physical causes of the motion.

Ironically, success only added to Kepler's frustration. Having caught a glimpse of what he might be able to accomplish with Tycho's data, he also became aware of how extensive the research project was that he had begun. Kepler concluded, "If I don't want to be deprived of the purpose of my travel, I have two choices. Either I copy his observations for my purpose (he is not going to accept that, and rightly, because these are the treasure he has spent his whole life and fortune on) or I find some way to stay and complete my own work." That meant trying to formalize his position at Benatky.

Kepler was to all intents and purposes Tycho's guest. Tycho had not regularized his position in any way, given him a contract, or even talked about long-term employment. The situation for Kepler's family back in Graz continued to be precarious. It was almost unthinkable that they should stay there, yet with no promise of regular financial support, Kepler could not drag Barbara away from her home and her extended family and ask her to forfeit her property. She would, Kepler knew, be even more likely than he to go mad in this clifftop beehive where men and women whose language she did not speak held all power and her husband did not even draw a salary. Kepler desperately needed a job commitment. He let Tycho know of his distress.

Tycho chose to interpret Kepler's requests for a contractual arrangement as a lack of trust in his honor and reliability. From Tycho's perspective there was, in fact, much to be said for letting this tense young man depart. He was not the good-natured, warmhearted man that Longomontanus was; he had had to be excused from the observation rota because of poor eyesight; his theories were eccentric; and he continued to lean toward Copernicanism. Furthermore, Tycho was not in a good position to make salary promises. Support from the treasury for the renovation was proving to be undependable, and Tycho had yet to see a gulden of the salary that Rudolph had promised him.

However, for Tycho, the moment of truth had arrived. He knew that validation of the Tychonic system could come only through skilled analysis of the observations, and Kepler had the potential to do the analysis. If Tycho wanted an authentic victory, and not merely the hollow, negative one of having no one able to gainsay him, he needed Kepler. Rising above the irritation of Kepler's complaints, he acted to try to remedy Kepler's problems. At the beginning of March, he wrote to Hoffmann, the man who had brought Kepler to Prague the previous January, suggesting that Hoffman meet him and Kepler to discuss Kepler's job position. While that meeting was being arranged, contract negotiations between Kepler and Tycho sputtered on, taking place through a variety of intermediaries, including Longomontanus. Kepler became more unhappy with each passing day. In his lifetime he endured many crises and indignities that would be sufficient to drive most men to distraction, yet only that spring at Benatky did his equanimity and good sense completely desert him. By his own report, he had a temper, but it had seldom been evident in public. In these tense days, he let it get the better of him several times at Tycho's dinner table.

Kepler grew so impatient, waiting for a reply from Hoffmann, that he took it on himself to lay out in writing the terms of employment he would accept. When Longomontanus suggested he moderate his demands, he submitted an only slightly revised memorandum. The demands were not unreasonable, but they were irksome to Tycho. They included permission for Kepler to go into Prague when he chose, though not to stay long; time off during the day for family affairs to compensate for having to work night hours; a guarantee of relief from observing duties, since his poor eyesight and lack of skill made him worthless at that anyway; Sundays and holidays off; and assurance that anything Tycho wished to publish under Kepler's name (such as an opinion of Ursus) would have to be at Tycho's expense and subject to Kepler's approval. Much more of a stumbling block were Kepler's demands that he have a salary both from Tycho and the emperor, that he be allowed to spend every day after noon on his own projects, and that he and his family be given a house separate from the castle. The house was particularly important to Kepler. "Tycho's house is very cramped," he wrote. "The turmoil of his family is great. I do not want to mix my family with them, because they are used to silence and modesty."

Feeling there was a strong chance that Kepler would leave, Tycho hedged his bets by extracting from him one more weapon against Ursus. Tycho chose Tengnagel to make the request that Kepler write an "opinion of Brahe's hypotheses, which a certain Nicolaus Ursus of Dithmarschen presumes to claim for himself." Though Kepler later reported that he wrote "Quarrel between Tycho and Ursus over Hypotheses," including a detailed history of the affair involving himself, Tycho, and Ursus, because Tengnagel "was eager to know these things better," the request almost certainly originated with Tycho. At the same time, Tycho assigned Longomontanus to write a similar piece refuting John Craig, an Aristotelian scholar whose debate with Tycho over the comet of 1577 still had not ended. Both Kepler and Longomontanus thought these assignments a foolish waste of time. It was likely that Kepler had his two-page document in mind when he listed among his demands the one having to do with items "published under Kepler's name." But Tycho was pleased with the result, and it was probably he who drafted the introductory statement including the words, "all this is compactly and rigorously refuted here by Kepler, an outstanding astronomer, thoroughly familiar with Ptolemy and Copernicus." He carefully filed away a copy of "Quarrel between Tycho and Ursus," together with a copy of the letter from

Kepler that Ursus had used in his book. Both, he thought, would represent important evidence in his postponed case against Ursus.

The contract discussion with Hoffmann never took place. Instead, on April 5 Kepler and Tycho met formally to discuss the matter of Kepler's employment, and Jan Jesensky, the emperor's physician, came from Prague to negotiate on Kepler's behalf. The three men sat down together, and Tycho brought out his response, in writing, to Kepler's demands. In return for the freedom to go to Prague, Tycho required a guarantee of silence about those aspects of the work at Benatky that Tycho preferred to keep secret. As for the request for Sundays and holidays off, that offended him, for he never asked his assistants to work on those days. He agreed to try to arrange separate housing for Kepler and his family, and he offered to pay the moving expenses, but he could not yet guarantee a salary. He had applied to the emperor and was waiting for a reply.

It appears that Tycho came to the bargaining table in good faith, ready to try to find ways of meeting Kepler's terms. He was fairly confident that there would eventually be some arrangement with the emperor, though he knew from prior experience the difficulty of actually collecting on such promises. Kepler, on the other hand, came to the bargaining table predisposed to be distrustful and argumentative. The meeting ended with angry words and no agreement. Later that day, at dinner, in the presence of the entire household, a frenzied Kepler, who had drunk too much wine, staged another outburst. This time, Tycho replied in kind. Kepler left Benatky the next day with Jesensky.

Tycho's equanimity following this episode was astounding, almost out of character, and perhaps stemmed only from the fact that Kepler had become so necessary for the realization of Tycho's ambitions. However, Tycho's response to others had always been more instinctive than reasoned, and in this instance both instinct *and* judgment must have told him that Kepler was not, either in respect to his talents or his character, an Ursus. Tycho refused to allow Kepler to de-

stroy himself. He treated Kepler with patience such as he had never
displayed except in matters involving those he most dearly loved in
his own family. Moving ahead in the assurance that all would be well
and that Kepler would come back, he informed Jesensky, before the
carriage departed for Prague, that he would need a written apology
from Kepler, and he sent a letter with Jesensky to Hoffmann, asking
for his intercession and help in smoothing things out.

The fact that Tycho did not react more strongly to his anger sent
Kepler into an even worse rage. The next day he penned such a blistering, insulting letter to Tycho that Tycho did not even have anyone
copy it for preservation, but sent it to Jesensky with a note saying he
regretted ever having anything to do with Kepler. This time he had
indeed been pushed too far. No one less than a king had *ever* gotten
away with treating him this badly. The note may have mentioned
Ursus, for Tycho requested that Jesensky "find out by a third or
fourth hand whether Kepler is quietly endeavoring to provide Ursus
too with some reproaches against me (for how shall I have confidence in him any more?)."

Somehow—perhaps Hoffmann did help in the end—Kepler came
to his senses, calmed down, and realized the enormity of what he had
done, and he wrote an apology to Tycho that was as effusively abject
as his former note had been vitriolic. Astonishingly, Tycho not only
immediately forgave him but climbed into his carriage and drove into
Prague (which he hated to do and avoided whenever possible) to
bring Kepler back himself. Shortly after that, they came to agreement
about Kepler's terms of employment.

Tycho's attempts to get an imperial stipend for Kepler soon met
with some success. The plan was that the emperor would summon
Kepler officially to assist Tycho for two years. During that time,
Kepler would continue to receive his salary as district mathematician
in Graz and in addition would receive a salary half that large directly
from the emperor. Though it seemed dubious whether Catholic
Styria would allow its Protestant mathematician that much leeway,

the hope was that since the assignment would come from the Catholic emperor of the Holy Roman Empire, of which Styria was a part, the Estates of Styria would feel sufficiently pressured to approve.

Kepler, his prospects looking much brighter, settled down to work and began making plans to move his family to Benatky. In May, Tycho arranged for him to travel the first leg of the journey to Graz with the Danish nobleman Frederick Rosenkrantz,* Tycho's third cousin who had fled Denmark when it was discovered he had got a lady-in-waiting pregnant. Rosenkrantz had been captured and sentenced to be stripped of his nobility and have two fingers cut off. Fortunately for him, that sentence had been commuted to service in the campaign against the Turks, the enemy whose earlier threat Kepler had successfully predicted in his first official yearly horoscope as district mathematician in Graz, and Rosenkrantz was on his way to join the army.

Tycho's arrangement for Kepler to ride as far as Vienna in Rosenkrantz's carriage may not have been entirely out of kindness: It was a good way to make sure Kepler got well beyond Prague without an opportunity to connect with Ursus.

---

*Kepler's traveling companion was immortalized in *Hamlet,* and would four hundred years later come even more into the spotlight in Tom Stoppard's *Rosencrantz and Guildenstern Are Dead.* William Shakespeare's "Guildenstern" was in real life Knud Gyldenstierne, another cousin of Tycho's. When the two kinsmen were in England on a diplomatic mission in 1592, they had made an impression on Shakespeare, although it must not have been a highly favorable one, judging by the way he portrayed them.

18

"LET ME NOT SEEM TO HAVE
LIVED IN VAIN"

1600—1601

WITH THE THAWING of the shipping lanes, in that same eventful spring of 1600, Tycho decided that the time had come to bring all his instruments to Benatky. The castle and the new bays along the bluff were almost ready for them. Tycho had left the four largest on Hven, and more recently, in order not to arrive so weighted down in Prague, he had stored most of the others in Magdeburg. During a year and a half their release had become entangled in bureaucratic red tape that not even a letter from the emperor had cut. Tycho hit on the strategy of suggesting to the Magdeburg authorities that they could make a tidy profit for themselves by transporting the instruments to Prague and using the return journey to import Bohemian wine. The instruments were almost immediately on their way.

Tycho had initially been much more worried about the four great instruments still on Hven, for he feared they would never be permitted to leave Denmark. The previous autumn, before Kepler arrived at Benatky, Tycho had sent his son for them and appealed for help to his brother Axel, now commander at Helsingborg. Longomontanus had left Rostock and traveled with young Tycho. The instruments did get out of Denmark and as far as Lübeck before winter weather

The Golden Griffin (far left) as it appears today.

made further transport impossible. With the spring thaw, they were on their way again through the mud and swollen rivers to Hamburg. Unaware that a bureaucratic delay there, similar to the one in Magdeburg, would halt them again, Tycho believed that before the summer was over he would have his new Uraniborg.

It was not to be. The plague had retreated. The court returned to Prague, and on June 10 Rudolph summoned Tycho to begin astrological consultations on state decisions. Because that would often require going to the palace twice a day, Rudolph arranged for accommodations for Tycho, his family, and his assistants in a hostelry known as the Sign of the Golden Griffin, in the section of the city where the palace stood. Tycho was now in precisely the situation he had hoped to avoid by choosing a castle away from the city.

Kepler's high hopes that late spring also came to a dead end. Not long after his arrival back in Graz, the Styrian councillors refused the request that they release him to work in Prague while continuing to pay his salary. There was, they argued, no useful purpose to which astronomical work could be put, and they suggested he go instead to Italy, study medicine, and return to practice as a physician. Since there could be no arrangement with Rudolph without the Graz salary, this decision ended Kepler's plans to return to Benatky. A despondent Kepler had no choice but to search for other possibilities. He even approached Archduke Ferdinand, suggesting that Ferdinand might, like his cousin the emperor, wish to have his own personal mathematician.

Ferdinand's reply, though not specifically targeting Kepler, was worse than a rejection. On July 27 a notice appeared, and the news spread rapidly: At 6:00 A.M. on July 31, all citizens of Graz had to present themselves for examination of their faith. Anyone who was not Catholic could pledge to convert. Those refusing so to pledge would be required to leave Graz.

Archduke Ferdinand himself was present with the commissioners as they sat at a large table in the middle of the church. It took three days for more than a thousand citizens to come to the table and be examined. Most were either already Catholic or agreed to conversion. When Kepler reached the table on August 2, he said he was a Lutheran, and he would not convert. His name was written down on the list of sixty-one banished citizens. He was given six weeks and three days to leave Graz for good.

The Kepler who had so completely lost his equanimity the previous spring at Benatky accepted this disaster with a serenity that even he found astonishing. "I would not have thought that it is so sweet," he wrote, "in companionship with some brothers, to suffer injury and indignity for the sake of religion, to abandon house, fields, friends and homeland. If it is this way with real martyrdom and with the surrender of life, and if the exultation is so much the greater, the greater the loss, then it is an easy matter also to die for faith." Kepler sensed that he was being forced to go, albeit not willingly, along a path he was meant to follow.

During this horrendous summer, when his family's whole world was falling apart, Kepler nevertheless found it much easier to work than he had at Benatky. Barbara seems to have found the strength to leave her husband and his calculations undisturbed in spite of the frightening and heartrending upheaval in their lives.

Kepler was trying to solve a problem having to do with optics that puzzled both Tycho and Mästlin. Tycho had recently observed a partial eclipse of the Sun by allowing the Sun's light to pass through a pinhole onto a white screen, where it made an image of the eclipsed Sun. Measuring this image, Tycho had concluded that the Moon was not large enough to cover the Sun completely, and that therefore a total eclipse of the Sun was impossible. Yet all astronomers knew that there were many total eclipses on record.

It seems that in spite of the lack of opportunities at Benatky for discussions, Tycho had spoken to Kepler about this problem and given him instructions on how to build a projection device for viewing an eclipse. Kepler constructed the device according to Tycho's pattern and observed a partial eclipse from Graz on July 10. Tycho had been right; the Moon's apparent diameter was smaller than the Sun's. Kepler proceeded to give serious mathematical consideration to pinhole images and whether this method of observing an eclipse distorted the image of the Sun. He concluded that the accuracy of the observation depended on the size of the pinhole, and that no pinhole could ever be small enough for complete accuracy. Hence Tycho's image of the Sun, distorted by the size of the pinhole he used, was slightly too large, and seemed incapable of ever being covered by the Moon. Before the summer was over, Kepler had defined the concept of "light rays" that still lies at the heart of geometrical optics. He wrote an essay about pinholes and light rays, and about his conclusions, and then set it aside when his personal difficulties became too pressing.

As soon as Kepler had learned there was no possibility of drawing a salary from Graz while he worked at Benatky or in Prague, he had written to inform Tycho that the hoped-for arrangement with the emperor would not be possible. Tycho's return letter came quickly: The failure of that arrangement must not matter, and he urged Kepler to come back, either with or without his family, "with confidence as rapidly as possible."

Though it was impossible to remain in Graz, Kepler had profound misgivings about putting his and his family's future at Tycho's mercy. Without a regular salary, living from day to day on handouts, even if Tycho were the most well-meaning and generous patron, was too precarious and demeaning an existence. Kepler wrote to Mästlin, asking whether some "little professorship" might not be found for him at Tübingen. Kepler knew Mästlin's reply could not reach him before he left Graz, so he told Mästlin to address it to him in Linz. Linz was on the way to Prague, and Kepler's plan was for Barbara and his stepdaughter Regina to travel that far with him and wait there while he went on alone to talk to Tycho.

It was two weeks beyond the deadline when, on September 30, Kepler, Barbara, and ten-year-old Regina left the city. They loaded two wagons with as many household goods and possessions as they could fit in and drove away from the comfortable home in the Stempfergasse where they had lived for three years. They had no real destination, and for Barbara the wrench was horrendous. All she had ever known or cared about, all her friends and extended family, all her quite considerable property, were here in Styria. Because of her husband's and her own religious convictions, they were leaving all that behind for an uncertain future among strangers. Barbara had an almost irrational fear of poverty, and there was no promise of a paying job ahead for her husband. It was unlikely they would ever be able to recover the value of her property.

On the journey north to Linz, they kept the hope alive that a letter from Mästlin would be waiting for them, but when they arrived in Linz, there was no letter. Kepler changed his mind about leaving Barbara and Regina alone, feeling it was better they all stay together, in case one of them became ill. They left the household goods behind in Linz to follow them later either to Prague or Tübingen—for there might yet be a letter from Mästlin—and continued toward Prague.

Kepler knew that now he was absolutely at Tycho's mercy—and not he alone but Barbara and Regina as well. He made one last attempt to bluff his way to a better job arrangement with Tycho, writing him in advance of their arrival, saying that because he had at one time received a scholarship from the duke of Württemberg, it was necessary that he pay his respects to the duke before settling elsewhere. While in Württemberg, he wrote, he would find out whether the duke would recommend him to another university, perhaps Wittenberg, Jena, or Leipzig, for his teachers at Tübingen had given him hope that the duke's connection with the university would be a great advantage—an extreme exaggeration on Kepler's part, if not an outright lie. His letter continued with the promise that if Tycho were to offer him an attractive position, he would give that offer first consideration. He named a deadline for the offer, four weeks at the latest. It was a feeble try, but it was the best Kepler could do.

Kepler's serenity had evaporated. To make matters worse, he developed a high fever on the journey from Linz to Prague. When the exhausted, depressed family arrived in the city, on October 19, 1600, it was the good Hoffmann who took them in, not Tycho. Although Kepler's fever was stubborn and would recur intermittently throughout the winter and spring of 1601, he was healthy enough to return to work for Tycho in late October. There had been no word from Mästlin, and no other possibilities had arisen. As Kepler would write, "God let me be bound with Tycho through an unalterable fate and did not let me be separated from him by the most oppressive hardships."

Kepler's fever was accompanied by a cough, and he thought he had contracted tuberculosis. Barbara also was ill and desperately unhappy, lonely in a foreign city, watching the little money they had disappear. She was not accustomed to enormous wealth, but she had always enjoyed a comfortable upper-middle-class standard of living. In Prague she would be reduced to maintaining their little family in a state of penury with nothing better in sight. The move had cost 120 gulden. Kepler's annual salary in Graz had been only 200 gulden. Even if his income had remained the same, everything was much more expensive in Prague, and now Kepler had no income.

Tycho made no "attractive offer" for Kepler to give "first consideration," and the four-week deadline passed. However, Tycho was trying to secure a salary for Kepler from the emperor. Rudolph, Tycho had reported, had given "a gracious nod," but both he and Kepler knew by now that even if that nod actually set the mechanism in motion by which a salary might eventually appear, a man and his family could starve waiting. For the time being, the Keplers had no choice but to depend on Tycho for everything. Tycho, who was already paying the Müller family's expenses out of his own pocket, because money for them from the emperor was overdue, now dug deeper and paid for the Keplers as well.

By the time the ailing Kepler was well enough to rejoin Tycho, Tycho had complained about cramped quarters at the Sign of the Golden Griffin, and the emperor had provided a private residence, not much larger. Into this house Tycho nevertheless managed to squeeze Kepler, Barbara, and Regina.

Eventually there must have been a salary agreement between Tycho and Kepler. The only record of it appeared when Tycho's daughter Elisabeth and Tengnagel were making plans to wed in the summer of 1601, and Tycho braced himself to pay for a lavish wedding. He told Kepler he would have to dole out Kepler's twenty-daler stipend in ten- and six-daler installments, because he did not have enough ready cash to do otherwise.

Meanwhile, in the autumn of 1600, Kepler, Barbara, and Regina still waited day after day in the faint hope that a letter from Mästlin would bring an offer from Tübingen. Mästlin's letter finally reached them in December. There was no job for Kepler in Tübingen, and Mästlin had no advice to give. "Here in Prague I have found everything uncertain," Kepler replied to him, "even my life. The only certainty is staying here until I get well or die." He continued to write letters to his mentor, pleading for help, but the old man would not reply again for four years. Evidently there *was* nothing he could do, except, as he had promised in that December letter, to "pray for you and yours."

Tycho's frustration and despair were almost equal to Kepler's. The months were passing with no possibility of going back to Benatky, and he abhorred the work Rudolph asked him to do. Though Tycho had not given up the idea that the movement of the planets and other celestial events somehow influenced life on Earth, he found astrological advice of the sort Rudolph wanted boring and a waste of valuable time. He was particularly ill at ease with the detailed predictions that would have best satisfied a monarch worried about military campaigns, the choice of generals, and the possibility of his own assassination (Rudolph had now reached the age at which his father had been assassinated). For Tycho, who believed that the free will of each human participant mitigated the influence of the stars, producing meaningful predictions about military campaigns, for example, seemed nothing short of ludicrous. Nevertheless, Rudolph was paying the bills, at least theoretically, and Rudolph believed devoutly in astrology. Tycho knew that were he to disabuse the emperor of that belief, he would quickly be out of a job.

The emperor also wanted advice that was more psychological and political than astrological. At court there were precious few who were politically neutral and could be expected to offer straightforward counsel without a personal agenda. Tycho's only agenda was getting back to Benatky with sufficient support to continue his work. Rudolph found Tycho's objectivity invaluable. Tycho thus had no choice but to try to meet Rudolph's needs and gear himself up to learn whatever he did not already know about dealing with competitors for imperial favor, secret alliances, opportunists hoping to link their careers to his own, and lies, exaggerations, and half-truths designed to thwart an ambition he did not even have

There was, however, more that needed to be dealt with than the usual affairs of the imperial court that summer and autumn of 1600. Not long after Rudolph summoned Tycho from Benatky, Rudolph suffered a temporary but severe mental collapse. It was not a rational ruler whom Tycho was advising. This was a particularly unfortunate moment for such a breakdown. Though Rudolph was emperor of the Holy Roman Empire, the area that fell under his most direct control was Bohemia, the northwestern part of what is now the Czech Republic—a multiethnic region ruled by a foreign dynasty (Rudolph's) that had not, even in the best of times, been free of explosive tensions. By the summer of 1600 this diffuse, smoldering enmity had become polarized by the Counter-Reformation and threatened to ignite in a major conflagration.

Rudolph II was a devout Catholic, but he opposed the more flagrant manifestations of the Counter-Reformation, not only in Bohemia but also in the larger empire. Incidents like the expulsion of Protestants from Graz and Styria represented a tragic failure in Rudolph's policy of attempting to keep Catholic zealots and Protestants, who actually were in the great majority, from all-out conflict. It was possibly in reaction to that failure in Graz that a mentally unstable Rudolph decided to expel a cloister of Capuchin monks from a residence where he had earlier invited them to live, near the palace. The normal Rudolph was not given to such unexplained acts. The monks accused Tycho of having influenced Rudolph to banish them because their prayers interfered with the black magic he was using to turn base metal into gold. (Had Erik Lange heard this accusation, he would have rushed to Tycho's side without delay.)

However much Tycho disliked the work he was doing, it is a tribute to his reawakened skills in the delicate handling of monarchs and the balanced nature of his counsel that he survived Rudolph's period of madness. Many other powerful men were permanently banished from court. Nor had Tycho contrived to remain on the periphery. He was considered one of Rudolph's closest advisers. Extensive correspondence survives in which moderate Catholic leaders communicated with him about influencing Rudolph to name his second brother Albrecht as his successor rather than his first brother Matthias, whom they judged to be virulently anti-Protestant.

On the more positive side, for Tycho, that summer and autumn, he finally received his own long-overdue salary. Also, there was undeniably a part of him that enjoyed moving in the most elite circles

at court, having powerful men trust in him and be aware of the emperor's enthusiasm for him. It was vindication for the treatment he
had received in Copenhagen. He also took pleasure in the company
of other well-educated people, of whom there were many in Prague.
Hoffmann had ordered a copy made of one of the quadrants in
Tycho's *Mechanica*, and the two men had used it to observe the same
solar eclipse Kepler observed from Graz in July.

For Tycho, another mitigating factor about the move from Benatky
to Prague was that he was able to initiate the long-delayed proceedings
against Ursus. Tycho had heard that Ursus was seriously ill. The various legal actions Tycho set in motion proceeded much too slowly and
inconclusively to satisfy him, for he chafed at the possibility that the
man he considered his nemesis, this slippery, underhanded swineherd,
would escape punishment by dying. In mid-August Ursus did just
that, with Tycho's lawyers harrying him even as he lay on his deathbed.
Ursus had not survived long enough to be, as Tycho reported that the
commissioners had promised him, "branded in infamy, and beheaded
or quartered according to Bohemian law."

Hence, by the time the Keplers arrived in Prague, Ursus in the
flesh was beyond Tycho's reach, but Ursus's book was not, and its
very existence was a threat. Tycho told Kepler that he was not so
much concerned with "destroying his person, whom everyone knows
was clownish and vainglorious, but rather his book, stuffed full of so
many insults and lies, and restoring the glory and reputation of myself and my associates." On order of the emperor, the printer sought
out all copies that could be found in Prague and consigned them to
the flames, and the book was banned throughout the empire. It was
an affront to Tycho when the council paid Ursus's widow three hundred gulden to compensate her for the confiscation of the books. But
Tycho could console himself that Rudolph in sound mind would
never have let that happen. It was a smaller setback than many suffered as a result of the emperor's brush with insanity.

With all the time spent carrying out the move from Benatky to

The Belvedere, a pavilion in the gardens of the imperial palace in Prague,
where Tycho set up his instruments in the autumn of 1600.

Prague, then from the hostelry to the house, while at the same time
dealing with a half-mad emperor, Tycho accomplished little meaningful work in the summer and autumn of 1600, and this situation
looked unlikely to improve. Longomontanus left, succumbing to the
homesickness that had been drawing Tycho's Danish assistants and
servants back to Denmark. Tycho reluctantly watched the departure
of this man who had helped him for so many years and joined him
in exile. Tycho had tried hiring various German scholars, none of
whom worked out successfully. More in need of analysis and computation than observation, he had been attempting to engage men
capable of that sort of work. But in the autumn of 1600 there was no
one else but Kepler in Tycho's employ with whose assistance he could
hope to complete his planetary theories in the way he wished.

The last of Tycho's instruments reached Prague in October, at
about the time Kepler himself returned. The emperor arranged for
Tycho to mount them on the balconies of an ornamental summer-
house in the palace grounds, now called the Belvedere. Tycho was
bitterly disappointed. He had hoped to use the need to install the instruments in their clifftop bays as an excuse to return to Benatky.

From the Belvedere's south-facing balconies, the emperor's palace complex blocked off a good portion of the southwestern sky.

By New Year 1601, Rudolf's mental state had improved, and his already high regard for Tycho had increased during the difficult months they had weathered together. When Tycho petitioned for citizenship and nobility for himself and his family in February, the emperor himself sponsored their petition. At last Kirsten and their children enjoyed a status that would permit them to inherit from Tycho and marry nobles of their new homeland. Tycho's estate was still large, especially if one included back pay from Rudolph (mounting up again), the value of instruments, books, and observations, and the loan (made shortly after he left Denmark) that he was finally calling in from the two young dukes of Mecklenburg. The future of Tycho's family at last seemed secure.

On the other hand, his scholarly future looked increasingly bleak. He had been near to having a new Uraniborg the previous spring. Now it seemed he would have to relinquish all hope of that, for Rudolph bought him the same palace he had rejected when he first came to Prague from Denmark. It was undeniably a beautiful house and garden, and an astronomer forced to live in the city could hardly have done better than this hillcrest location. Nevertheless, the mansion still had the same disadvantages that had caused him to turn it down eighteen months earlier: The tower was not large enough, and the location was too accessible to the court, just a few minutes' walk west of the imperial palace and no time at all in a carriage.

Again, Tycho's enjoyment of a splendid, nearly royal lifestyle was an antidote to despair. He moved his library into the house, the three thousand books that had been waiting all this time in Magdeburg. He took smug satisfaction, when spring came, in mentioning in letters inviting former acquaintances and relatives in Denmark to his daughter Elisabeth's wedding that the summer nuptials would be held in his palace, formerly belonging to the vice chancellor of the empire. The contingent from Denmark was not expected to make an appearance, but it was a triumph to inform them how luxurious and pampered his present situation was and that his daughter was marrying a nobleman.

Nevertheless, the discouragement of having to move family and research establishment again, not back to Benatky but to this unwanted house, was a serious drain on the fifty-four-year-old Tycho's energies. During that winter, his friends began to notice that he had lost some of his usual spark and seemed to be resigning himself to old age and declining health. Tycho's brother Jørgen, who was much younger than he, died in February. Jesensky reported that in the middle of a cheerful conversation Tycho would change the subject to talk about death. Kepler commented in a letter to Mästlin, who was still not answering, that Tycho was acting childish and capricious, though he was "still good-natured," and that Tycho seemed burdened with cares: "He always resembles a lost man, but always somehow extricates himself. His success at this is to be wondered at." Tycho made some progress reorienting his instruments for their new location, but he did little observing and failed to move ahead at all with the books that he had been working on for many years. Several of these were near completion, and it would not have required a great deal of effort to finish them, but he lacked the energy and interest.

When Tycho and his family moved to the mansion, the Keplers moved there too. Yet in spite of the difficulty Tycho was having finding good computational help, he wasted Kepler's talents that winter, partly because Kepler wasn't at his best—his fever kept returning—but also because Tycho was still paranoid about his observations and not satisfied that he had completely defeated Ursus, even though the man was dead and all known copies of his offensive book had been destroyed. Kepler complained later in a letter to the astronomer Giovanni Antonio Magini that Tycho would show him his "choicest" observations, but only "inside his four walls," and say to him, "Get to work." If Kepler asked to see observations other than those Tycho set before him, Kepler was told he was being too inquisitive. "If only I could copy them quickly enough!" Kepler wrote to Mästlin, and in the same letter mentioned an idea for prising some of the observations out of Tycho: "If you would send him some of your observations, he would, I think send some to you, too, if you ask him to do so. For in spite of all the instability of his character, he is, after all, a man of great benevolence." Mästlin did not reply.

As for Kepler's own astronomy, he spent a little time on some theories about Mercury, Venus, and Mars, discussed them with Tycho, and thought about the orbit of the Moon, but this was not a productive winter. "A fever gripped me," Kepler later recalled. "In the meantime I wrote against Ursus on Tycho's orders." In a letter to Mästlin he complained, "Because of this illness of mine I am doing nothing but write against Ursus." Kepler found it distasteful to carry on the dispute after Ursus's death, but Tycho was still obsessed with proving that he, not Ursus, had invented the Tychonic system. Not only did he want to destroy Ursus's scientific credibility, to keep him from ever getting credit for it, but he also wanted to discourage others who, he believed, were also guilty of plagiarizing his system. The main current target of Tycho's fears was a Scotsman named Duncan Liddell, whom Tycho had suspected ever since Liddell visited Hven in 1587 and 1588. Liddell seems to have been completely innocent; over the years he remained a responsible scholar and teacher and one of Tycho's staunchest supporters, though he kept his distance because of Tycho's suspicions and hostility.

Kepler put his feverish head to the task of coming up with something that would satisfy Tycho's instructions that he "rebut even more clearly and more fully than you have done previously Ursus's distorted and dishonest objections to my invention of the new hypothesis . . . and ascribe the new hypothesis to me, as is right, just as you did before with demonstrable reasoning." Kepler did not finish his "Defense of Tycho against Ursus" that winter and spring. He would resume work on it several years later, but the still unfinished manuscript was not published until 1858. Kepler made the most of a poor assignment. It is one of the finest analyses ever written about scientific methodology, pointing out a difference between the Ptolemaic and Copernican models that was of profound importance to Kepler and that remains even today the primary reason for deciding in favor of Copernicus. In principle, Ptolemaic astronomy was not "incorrect." It could plot and predict the courses of the heavenly bodies just as correctly as Copernican astronomy. So could the Tychonic model. But, wrote Kepler, "If in their geometrical conclusions two hypotheses coincide, nevertheless in physics each will have its own peculiar additional consequence." In other words, when one began asking the "why" questions, seeking the physical causes for the motion, Ptolemaic and Tychonic astronomy could no longer hold their own. To Kepler, the search for physical causes had become paramount.

In April, with his health not improved, Kepler interrupted work on this treatise to make another trip back to Graz. Barbara's father had died, and Kepler needed to salvage whatever he could of his wife's inheritance. Most of that was tied to estates and useless to the Keplers unless it could be converted into cash. Though the Graz authorities did not block Kepler's return, in a financial sense the four-month trip was an exercise in futility. Nevertheless, Kepler finally shook off the fever that had afflicted him for nearly a year, and he wrote to Barbara that he was enjoying visiting friends, who everywhere were treating him as a welcome guest.

Barbara wrote that she was not getting as much money from Tycho as he had promised. She could not buy wood for the fire. An angry exchange of letters ensued. Tycho told Kepler to calculate what was owed and he would be paid, but to behave in future more considerately toward his "benefactor" and "have more confidence in him." Kepler bristled at this insinuation that Tycho was giving him charity instead of fair recompense for his work. The contretemps finally ended agreeably, but it was symptomatic of the dissatisfaction Kepler still felt with his working arrangement.

Kepler kept up with events in Prague through Barbara's letters, partly written in the Keplers' own secret code, which would not have helped Tycho's paranoia if he had known about it. The most important occasion of the summer was the wedding of Tycho's daughter Elisabeth to Tengnagel in June. Kepler inquired of Barbara in code whether the bride looked pregnant. Perhaps she did, for Tycho's grandson was born in late September.

Though Tycho expressed some displeasure with the young couple prior to the wedding, perhaps because of the pregnancy, he was immensely gratified that his daughter, who in Denmark was not even considered his legitimate child and could never have married into the nobility, was marrying Tengnagel. Although Tycho had previously referred to him as his *domesticus,* or servant, Tengnagel was a nobleman, a man of great political promise whom Tycho had known and trusted for many years. In this marriage Tycho and his family were repaid a little for the grief and disgrace of Magdalene's ill-fated betrothal to Gellius. Tycho did everything in his power to make the triumph as public as that embarrassment had been, and the invitation list was long and illustrious. Even Rudolph was invited, though there was no expectation that the reclusive emperor would attend. Tycho's sister Sophie, who had been intimately involved in Magdalene's sorrows, hoped to come, but ill health prevented her at the last minute. She had previously begun several journeys to Prague and had to interrupt them midway because of Erik Lange's recurring disasters. Elisabeth and Tengnagel left after the wedding for the Netherlands along with another of Tycho's assistants, Johannes Erikson, so when Kepler returned in late August, a healthier, more optimistic, though poorer man, the house by the wall was less crowded.

By that autumn of 1601 the decision Tycho had been trying to make for nearly a year and a half had taken on greater urgency. His work was not finished and would not be, in his eyes, until there was complete justification of his belief in the Tychonic system. His astronomy would continue to languish, incomplete, with no hope of accomplishing what he had spent a lifetime working toward, unless he finally put his full trust in Kepler. If Tycho was indeed burdened with premonitions that he had not long to live, he surely anticipated that that decision would mean trusting Kepler beyond his death. And—though

Tycho had to have recognized that Kepler would put the observations to better use than any other man available—trusting Kepler could not mean trusting him ultimately to support the Tychonic system. Nevertheless, with Ursus buried for more than a year, Tycho at last made the leap of faith and wagered his earthly immortality on Kepler.

Tycho invited Kepler to accompany him to the imperial court. There, for the first time, he introduced him to Emperor Rudolph and proceeded to make a dramatic proposal: Tycho and Kepler would take on the prodigious task of compiling a superb new set of astronomical tables based on Tycho's observations and more accurate than any the world had ever known. With the emperor's gracious permission, these would be named the Rudolfine Tables. Great astronomical tables in the past, such as the Alfonsine Tables, bore the names of their royal sponsors. The Rudolfine Tables would, similarly, be a monument to Rudolph and a testament to his generous support of learning. There was nothing further required of the emperor than what he had already granted Tycho . . . except for a salary for Johannes Kepler.

The emperor was thrilled with the idea. The paperwork for Kepler's salary was begun posthaste.

Tycho had previously provided for the future of his family. Now he had also provided for Kepler's future. With this bold decision, Tycho placed all his precious observations in Kepler's hands. Tycho's secrets would no longer be secrets from Kepler.

❧

ON OCTOBER 13, 1601, only a few days after meeting with the emperor, Tycho accompanied a friend, Councillor Minckwicz, to dinner at the palace of Peter Vok Ursinus Rozmberk just a few steps from the emperor's gate. Courtesy forbade one to rise from the table for any reason before one's host had risen, and it was Tycho's adherence to this simple point of etiquette, "so trivial an offense," as he himself put it, that brought him to his deathbed.

All the barriers that had kept Tycho and Kepler at a distance had now disappeared, and it is Kepler's description that provides more intimate details about the days that followed than are available about any other episode of Tycho's life. Kepler wrote,

Holding his urine longer than was his habit, Brahe remained seated. Although he drank a little overgenerously and experienced pressure on his bladder, he felt less concern for the state of his health than for etiquette. By the time he returned home, he could not urinate any more. [Kepler here noted down the positions of the Moon, Saturn, and Mars on the night of the banquet.]

Tycho's own medical expertise was considerable, and he tried various remedies, but with no success. He endured five days and nights of agony, unable to sleep.

Finally, with the most excruciating pain, he barely passed some urine. But, yet, it was blocked. Uninterrupted insomnia followed; intestinal fever; and little by little, delirium. His poor condition was made worse by his way of eating, from which he could not be deterred. On 24 October, when his delirium had subsided for a few hours, amid the prayers, tears, and efforts of his family to console him, his strength failed and he passed away very peacefully.

At this time, then, his series of heavenly observations was interrupted, and the observations of thirty-eight years came to an end. During his last night, through the delirium in which everything was very pleasant, like a composer creating a song, Brahe repeated these words over and over again: "Let me not seem to have lived in vain."

It was only to God and Kepler that the prayer could have been addressed. Several years later, Kepler added to his description in a

Tycho Brahe's tomb in the Maria Tein church in Prague.

chapter of his book *Astronomia Nova* that when Tycho lay dying, "although he knew that I was of the Copernican persuasion, he asked me to present all my demonstrations in conformity with his hypothesis."

Perhaps, had there been a choice, Tycho would have preferred burial on his once-beloved Hven, or in the parish church in Kågeröd on his ancestral estate where his parents and other family lay. In Prague, even at Benatky, he had always felt like a man far from home. But it is doubtful that Denmark, even had it chosen to do so, could have given Tycho a more magnificent burial than Prague did, or one that would have pleased him more. Kepler described it:

The casket was draped with black cloth and decorated in gold with the Brahe coat of arms. In front of the casket were carried candlesticks, likewise adorned with his arms, and a black damask banner displaying his titles and arms, in gold. Behind the casket was led his riding horse, followed by a black taffeta banner and then another horse draped in black cloth. [Followed by men walking single file, carrying Tycho's sword and armor.] The casket was borne by twelve imperial officials, all noblemen. Behind the casket walked Tycho's younger son, between the Swedish count Erik Brahe and Baron Ernfried von Minckwicz, in long mourning dress. They were followed by other imperial councillors, barons, and noblemen, Tycho's assistants and servants, then Tycho's wife, guided by two distinguished old royal judges, and finally his three daughters, one after the other, each escorted by two noble gentlemen. Then proceeded many stately women and girls, and after them the most distinguished citizens. The chairs in the church in which the family sat all were draped with black English cloth. The streets were so full of people that those in the procession walked as if between two walls, and the church was so crowded with both nobles and commoners that one could scarcely find room in it. When the sermon was over, the banners, helmet, shields, and other arms were hung over the crypt.

Kepler did not record where he walked in the procession. He must have been among the assistants and servants. However, he already knew that he was no longer to continue as any sort of "hired hand."

19

THE BEST OF TIMES

1601—1606

KEPLER HAD THE NEWS two days after Tycho's death: Barvitius, the imperial secretary, came to tell him that the emperor had named him imperial mathematician, and he should apply for a salary immediately. The title carried with it responsibility for the care of Tycho's instruments and manuscripts, as well as for the completion of his unfinished work, most urgently the Rudolfine Tables. The legacy was Kepler's. Tycho's full set of observations had fallen into his hands, the pages open at last.

Tycho's instruments and intellectual property were not really the emperor's to bestow, for they belonged to Tycho's family. So Rudolph purchased them for 20,000 florins, more than Kepler at his old Graz salary would have earned in a century. Of course, as the Brahe family were aware by now, collecting on such a promise from the emperor was no easy matter. To Kepler, the value of the observations was beyond any price.

Following Tycho's death and burial, the Keplers moved out of Tycho's mansion by the wall to a house across the river from the imperial enclave, in a section of Prague known as the New Town (it dated only from the fourteenth century). For the first time since leaving Graz, they had a home of their own. Their house was across the street from the Emaus cloister, an hour's walk from the palace. Tycho would have been pleased to have that much distance between himself and the emperor, though it was a long journey for Kepler when he had to make it.

The first decade of the seventeenth century was a glorious time to be living in Prague, albeit an expensive one. With the court in residence, it was the center of political life in Europe, wealthy, cosmopolitan, rich in history but moving with energy into the new century. Its narrow streets echoed with many languages. Kepler called it "a gathering of nations." Carriages of courtiers drove along its wider avenues and stopped at splendid houses that were not only impressively large but also exquisite in their proportions and details and furnished with treasures. Other men than Tycho and Kepler also learned that the royal coffers were unable to make good on all the promises Rudolph II made. Nevertheless, Rudolph's interest in learning and the arts set the tone of a court and a wider community that drew many superbly productive artists and scholars to Prague. One of Rudolph's numerous idiosyncracies was a shyness that at this time in his life caused him to shut himself away for days at a time, paying little if any attention to the activities he supposedly supported, but his name and reign are still linked with a great flowering of the arts and scholarship.

There was no higher honor to which an astronomer could aspire than the one that was now Kepler's. After so much despair and struggle, it seemed that he and his family were at last on their feet. Barbara Kepler had a house of her own to manage, and in July 1602 she gave birth to a daughter, Susanna. The little family that had consisted only of three uncomfortable refugees—Johannes, Barbara, and Regina—was growing larger and looking forward to happiness and prosperity, with freedom from religious persecution.

The years he lived in Prague were indeed golden years for Kepler, the peak years of his life, with many friendships, respect he richly deserved, and splendid scientific accomplishments, but the difficulty in collecting his salary cast a pall over them. Not until five months after Kepler's appointment did he receive the first payment, and he continued to encounter obstacles collecting even a pitifully small portion of what he was owed. He made such a pest of himself with the royal treasury that a nasty note about him is still appended to the treasury records. When persistence failed, and it almost always did, the Keplers fell back on meager revenues trickling in from Barbara's property and whatever extra compensation Johannes could scrape up here and there. The Keplers did not, however, live in poverty. Their home was "simply run," as Kepler put it, but the lifestyle they managed to maintain was comfortable and appropriate for a man in his position. Kepler's wardrobe included fashionable attire with a standing lace collar, the expected work clothing at court and when an imperial mathematician appeared in public.

Barbara Kepler did not find Prague nearly so congenial as her husband did. What is known about her comes almost entirely from Kepler's letters, and in these when he mentioned her he most often wrote in her defense, placing the blame on himself that their marriage was not happy. Though their lifestyle in Prague was similar to the way Barbara had lived when she was a young woman, she had few skills in simple household economy and, to judge from a few of Kepler's more candid letters, opted instead for a miserliness perhaps born of fear of sinking into true poverty. Whether her overzealous attempts to economize were well-meant self-sacrifice or a kind of self-martyrdom thrown in the face of her husband is uncertain. She chose, for instance, to cut back severely on her own clothing budget, at the risk of becoming an embarrassment, in order to spend everything on her children.

While Kepler flourished, Barbara grew melancholy and was bitter about real and imagined differences between her life and the lives of the women she saw around her. Her husband wrote that she had neither "the heart nor the means" to make herself better known in Prague society, a plight she may have lamented as much as he. She did, however, make a good impression on some who met her. Contemporary descriptions call her lovely in appearance, polite, respectable, modest, pious, and generous toward the poor.

At home, where she was burdened with unpleasant economies and a husband often buried in his studies, the more sympathetic Barbara seems not to have been much in evidence. In one of his later letters Kepler described Barbara as "weak, annoying, solitary, melancholic" and "fat, confused, and simple-minded." She immersed herself in prayer books, yet for all her piety she could not curb an ugly temper. Kepler also was not unfailingly placid and long-suffering, and of their relationship he admitted, "There was much biting and getting angry, but it never came to any hostility . . . both of us well knew how our hearts felt toward each other." The quarrels would end when Kepler saw that something he had said had deeply hurt Barbara. Overcome with guilt, he would stop immediately. But "not much love befell" him.

Barbara did not understand astronomy, and though she had followed him into exile because of his conscience, and both of them were deeply religious, there was little if any discussion of religion between them. Kepler, either because he thought his mature, complicated faith would disturb his wife's simpler beliefs, or because these were not matters to discuss with a woman, always spoke in Latin and avoided German (the language she understood) when he conversed about religion with visitors to their home.

Meanwhile, though Kepler may have felt he lacked for love from Barbara, he was deeply and widely loved in Prague. Both as imperial mathematician and as a private individual, he received attention and appreciation of a sort he had never known before and had always, shamefacedly, longed for. The emperor himself kept up with Kepler's scientific work, and visiting dignitaries and royalty sought his company. He had many devoted personal friends, from the highest court officials to simple uneducated people whose uninformed opinions

about astronomy and astrology he seemed genuinely to value as a
spur to his own thinking. His old friend Hoffmann, who had first
brought him to Prague, remained close and provided him with two
astronomical instruments, for Tycho's instruments remained locked
away, unavailable, waiting for the emperor to make good on the
promised payment to the Brahes. Kepler also kept up a lively correspondence—for he was an engaging letter writer—with numerous
acquaintances and scholars.

As imperial mathematician, Kepler was expected to produce calendars—as he had in Graz—and to give Rudolph advice based on astrology. Kepler had become less and less fond of casting horoscopes,
an activity he now described as "unpleasant" and "begrimed" work,
which should nevertheless not be "smothered." He began applying a
new, more scientific approach, attempting to trace whatever appeared to be established from experience back to causes and physical
links, and in 1605 he stopped producing prognostications entirely.
His advice to Rudolph often came in the form of essays, not all directly related to astrology, for Rudolph had got into the habit of asking Kepler's predecessor Tycho for many kinds of advice. One essay
was an opinion about a dispute between the Republic of Venice and
Pope Paul V about a pump without valves that Kepler had invented.
Another had to do with Galileo's discoveries with the telescope.

When it came to Kepler's scholarly work, it was antagonism with
the Brahe family that unexpectedly determined his research agenda.
At the time of Tycho's death and funeral, Tengnagel, Elisabeth, and
Tycho's eldest son were away. Kirsten was distraught with grief, and
of the rest of the family only Magdalene and Tycho's younger son
Georg were in Prague. Kepler did not wait to consult them or find
out how financial matters would be settled. He took charge of
Tycho's astronomical observations. During the year after Tycho's
death, he reveled in the freedom to consult them whenever he
wished, without Tycho to snatch them away and accuse him of being too inquisitive.

Tengnagel returned to Prague in October 1602 and found that there had been almost no payment from the royal treasury for the instruments, observational logs, and manuscripts. He also could not discern that Kepler had made any progress on completing the manuscripts. Through machinations at court, Tengnagel contrived to have the task of composing the Rudolfine Tables transferred to himself, at double Kepler's salary, though Kepler continued to be imperial mathematician. To Kepler's way of thinking, he had made plenty of progress, and it was inconceivable that he should relinquish the Mars data when he was so close to answers. He handed over most of the material, but not the Mars observations. Those he secretly kept, thinking it improbable that Tengnagel would actually consult the observations himself and notice that something was missing.

In the course of this unpleasantness between Kepler and Tengnagel, the emperor in the autumn of 1602 inquired of Kepler what he planned to publish in the near future to justify his employment as imperial mathematician. Kepler made some rapid decisions. Taking stock of his unfinished works, he promised two books: First, within eight weeks, by Christmas, he would complete *Astronomiae Pars Optica* (The Optical Part of Astronomy). That, he thought, was nearly finished already, the fruit of the summer he had spent in Graz prior to his family's expulsion. Second, he informed the emperor, by the next Easter (1603) he would complete "Commentaries on the Theory of Mars." He had been working on Mars, albeit with many interruptions, since he had first joined Tycho at Benatky.

Kepler was being overly optimistic about completing *Astronomiae Pars Optica* by Christmas. He was still the same man who at age twenty-six had written that his eagerness led him "to think of a lot of things as easy, which proved difficult and time-consuming in the carrying out" and who "in writing [would] continually start thinking about new things." He had already begun to consider many elements of optics that were relevant to astronomy besides those he had been investigating in Graz—for instance, the extent to which light is refracted as it enters the atmosphere, the question that had plagued Tycho. Kepler put his mind to this problem without complete success, for he used erroneous data. Also, he decided there should be exhaustive treatment in the book of eclipses and the sizes and distances of the Sun and Moon. He anticipated no great problem there, for he was currently producing a treatise on the subject. Soon he decided, however, to keep that material separate and not include it. But he needed to delve into the function of the human eye.

In this area, Kepler met great success. The question of how the eye works was not new, and there were theories that attempted to explain it, but Kepler's previous optical analyses gave him the background to discover, after rigorous calculation, that the old theories were wrong. Applying his idea of light rays, he was the first to realize that the image of the outside world is not captured in the fluid of the eyeball. It is, rather, projected by a lens in the eye onto the surface of the retina. Working like a "pencil of light," as he called it, the light rays "draw" the image on the retina. Kepler discovered that the image is upside down and backward on the retina. He was not able to explain how the mind compensates for this, but he did arrive at a precise understanding of the way in which differently shaped eyeglasses could correct nearsightedness and farsightedness. This was a particularly relevant question for him, because he wore spectacles himself.

In the introduction to the book, Kepler spoke of another discovery he had made: the inverse square law of light. If a burning candle is set on a table, the lighted area surrounding it on the table is a circle, with the candle in the center of the circle. Kepler, thinking in three dimensions rather than two, reasoned that light, starting from one point in space (the candle flame), spreads out not just in a circle but in *all* directions, in the form of a sphere. Wherever you are, within reach of the light, you can think of yourself as being at the edge of a sphere centered on the light source. Someone a little farther away can also imagine himself or herself being at the edge of a sphere centered on the light source. At that second location, the sphere is bigger, and the light looks dimmer. How much dimmer? was the question. Kepler reasoned that the light's brightness was related to the size (the area) of the sphere. If two observers were both looking at the light, and observer B was twice as far away as observer A, then observer B's sphere was four times as large as observer A's sphere. B saw the light only a fourth as bright as A did. If B was three times as far away as A, B's sphere was nine times as large as A's sphere, and B saw the light only a ninth as bright as A did. As Kepler's inverse square law of light states, the intensity of light is inversely proportional to the square of the distance. The square of two (two times as far away from the light source) is four. The square of three (three times as far from the light source) is nine, and so forth.

In January 1604, a little more than a year after the Christmas for which he had promised it, Kepler presented the completed manuscript of *Astronomiae Pars Optica* to the emperor, and the book went into publication. The ideas and discoveries about light and optics that Kepler wrote about in this book and later applied to the telescope in another work, *Dioptrice*, became the foundation for seventeenth-century optical theory.

After their clash in the autumn of 1602, Kepler and Tengnagel were often at loggerheads, but they managed to make some joint progress on the completion and publication of Tycho's posthumous works. There had been an extremely uncomfortable moment in the spring of 1603 when Tengnagel, turning his attention to the Rudolfine Tables, had discovered that the Mars observations were missing. Kepler reluctantly surrendered them. He was at that time writing *Astronomiae Pars Optica*, and the removal of the possibility of working on Mars may have caused him to go as deeply as he did into that other subject.

Tengnagel's most impressive talents lay in politics and diplomacy, which Tycho had recognized and put to use when courting favor with the royalty of Europe. Hence more and more of Tengnagel's time was taken up with Hapsburg politics and the deteriorating political situation in Bohemia and Germany. In the summer of 1604 he conceded that he could not possibly complete the Rudolfine Tables by himself. In return for a promise to complete them in a manner satisfactory to Tengnagel and to seek his approval before publishing anything based on the manuscripts, Tengnagel allowed Kepler to use some of Tycho's observational journals.

The precious Mars observations were once more in Kepler's hands. However, promises to Tengnagel notwithstanding, he did not set to work on the Rudolfine Tables. He was again engrossed in his study of Mars's orbit. His book about that, promised for Easter 1603, was well behind schedule. It would, in fact, not be ready to go to print until late in 1605, and it would have a new and highly appropriate title, *Astronomia Nova*—New Astronomy.

In the autumn of 1604 the Keplers, with Barbara six months pregnant, moved house again, this time to Wenzel College in the Old Town, nearer the palace but still across the river. One of Kepler's dearest friends, Martin Bachazek, rector of the University of Prague, lived there. Kepler relished the opportunity to converse with him daily.

That same autumn, not long after the move and while Kepler was wrestling with one of the most difficult problems in his book about Mars, a celestial event occurred that left him no choice but to abandon his writing desk and his columns of calculations. The event began inauspiciously when a court official in a state of great agitation roused the family at dawn on October 11. When Kepler could make sense of the man, he learned that on the previous evening he had seen a brilliant new star through a gap in the clouds. Kepler was skeptical. Six days of overcast skies followed. He had almost forgotten the incident when on the evening of October 17 the sky cleared. He saw the star himself and realized that the messenger's excitement had been justified. As bright as Jupiter, sparkling like a diamond in all the colors of the rainbow, this nova appeared in the sky near Saturn and Jupiter, which were near conjunction. Mars was also close by. Tycho had had his "star," and now Kepler had his.

As Kepler's fateful drawing for his class in Graz had demonstrated, Jupiter and Saturn come into conjunction every twenty years. The regular pattern in which the conjunctions occur means that ten conjunctions happen within each of four areas of the zodiac. Astrologers associate the areas with the four elements identified by Aristotle: fire, water, earth, air (see figure 12.1). It takes eight hundred years for the conjunctions to pass through all four areas, known as "trigons." The conjunction of Jupiter and Saturn at the time of the appearance of "Kepler's star" marked the beginning of the two-hundred-year period in which the conjunctions would occur in the trigon associated with the element fire, the "fiery trigon." Any conjunction was considered to have important effects on human events, mostly bad; a conjunction in the fiery trigon presaged even greater calamity.

In view of the astrological implications of such a conjunction *and* a new star at the same time, Rudolph would not rest until he and all the other nervous citizens of the empire could be informed by the imperial mathematician what they should make of this wonder. Bachazek built a small wooden tower so that Kepler and he could see the star better, and Kepler almost immediately produced a delightful short report to reassure the emperor and the populace of Prague. Among other possibilities, Kepler predicted (with tongue in cheek) good sales for booksellers, because every theologian, philosopher, physician, mathematician, and scholar would want to publish his own ideas on the matter.

To help fulfill that prediction, two years later Kepler himself dedicated a book about the nova to the emperor, based on research and continual observation as it faded. *De Stella Nova*'s subtitle was *A Book Full of Astronomical, Physical, Metaphysical, Meteorological and Astrological Discussions, Glorious and Unusual.* There was a widely held opinion that the planets had ignited the nova. Kepler insisted it was much farther away than the planets, at the distance of the fixed stars, and he made a good case (based on erroneous data) that the fixed stars were not suns. He also rejected the suggestion that a group of atoms had come together by pure chance to form a new star. That, he wrote, was like thinking that "if a pewter dish, leaves of lettuce, grains of salt, drops of water, vinegar, oil and slices of egg had been flying around in the air for all eternity, it might at last happen by chance that a salad would result." He even had mentioned the matter to his wife as she set a salad on the table before him. As he wrote: "'Yes,' responded my dear, 'but not so nice as this one of mine.'"

Kepler ruminated a bit about the astrological implications of the star, but he ended by telling his readers that the best advice he could give them was to examine their sins and repent. Star or no star, that could certainly do them no harm.

Modern research shows that Kepler's Star, like Tycho's in 1572, was a Type 1 supernova. There have been three in our galaxy in the last thousand years. (The other was in 1006.) Kepler's was the last supernova that would be visible to the naked eye until 1987, when one occurred in the Large Magellanic Cloud, a satellite galaxy of the Milky Way. Tycho and Kepler had no idea how extraordinarily fortunate they were each to see one. The serendipitous salad was perhaps not so much less likely after all.

The Kepler family kept growing. In early December 1604, Barbara gave birth to a son, Friedrich, who was to be a great favorite of Kepler's. The domestic disruption surrounding his birth caused Kepler to exclaim in a letter, "For what a business, what an activity, does it not make to invite fifteen or sixteen women to visit my wife, who lies in childbed, to receive them hospitably, to see them out!" Perhaps he should have reconsidered his lament that Barbara had neither heart nor means to make herself further known in society. Sadly, as Barbara's house became livelier with children, she herself withdrew further into melancholy.

For Kepler the astronomer, 1604 was both a frustrating and an exhilarating year as he struggled daily to solve the riddle of Mars's orbit. He came to think of it as a war with Mars, by ancient tradition the most warlike of the planets. By the time Friedrich was born, Kepler still wasn't sure whether he was nearing success, whether victory might still be several years off, or whether the orbit of Mars was perhaps not mathematically describable at all. *Astronomia Nova* was not a report on the results of completed research. Kepler had written fifty-eight chapters of the book in almost final form before he discovered that Mars's orbit is elliptical.

Tycho's records contained plenty of data on Mars, for the "problem of Mars" had caused him over a long period of time to make that planet the focus of many observations. That "problem" was the catalyst that led Kepler to his first two laws of planetary motion.

Already in antiquity, observations of Mars had made it clear that the speed of the planet does not remain uniform throughout its orbit. Astronomers in the intervening centuries had used ingenious devices to describe such irregularities in a mathematical/geometrical way. Once such device was an "eccentric" orbit—an orbit not precisely centered on the center of the system (not precisely on Earth for Ptolemy; not precisely on the Sun for Copernicus). A straight line drawn through the center of the system (Earth or Sun) and the center of the eccentric orbit was called the apsidal line. Extended farther, the apsidal line passed through the point where the planet was farthest from the center of the system (at aphelion) and the point where it was closest (at perihelion) (see figure 19.1).

Common experience indicates that objects *seem* to move much more slowly the farther away they are. A bird flying close overhead can easily appear to win a race with a plane far higher above it in the sky. Closer *looks* faster. Ptolemaic astronomers, Copernicus, and Tycho had all considered the possibility that what appears to be a variation in the speed of a planet, as viewed from the center of the system (Earth or Sun, depending on whether one followed Ptolemy or Copernicus), is only an illusion created by the fact that the planet, traveling on an eccentric orbit, is sometimes closer than at other times. But they had realized that this explanation was not sufficient

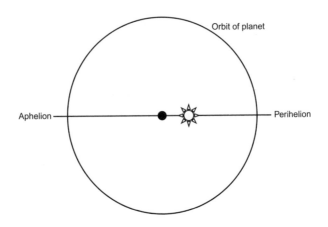

Figure 19.1 (Apsidal line): An off-center orbit was said to be "eccentric." The distance between the center of the orbit and the center of the system (Earth or Sun) was known as the "eccentricity" of the orbit. A straight line drawn through both of those two points is the apsidal line. Extended farther, the apsidal line passes through the point where the orbiting planet is farthest from the center of the system (at aphelion) and the point where it is closest (at perihelion).

to account for the extent of speeding up and slowing down revealed by observations. Ptolemy experimented with an equalizing point or "equant" that was not the center of the system (Earth in his case), and not the center of the eccentric orbit. It was a third point along the apsidal line, a point from which an observer would find that the planet appeared to be moving with constant speed.

The "problem of Mars" was that it was the most difficult of the three outer planets to accommodate in this manner. Mars has by far the greatest eccentricity. Tycho must have been well aware, when he decided to train the instruments of Uraniborg and Stjerneborg on Mars, that that planet provided not only the stickiest problems but also the best opportunity to define what the problems were and, one might hope, to solve them. As Kepler put it, "I consider it a divine decree that I came at exactly the time when Longomontanus was busy with Mars. Because assuredly either through it we arrive at the knowledge of the secrets of astronomy or else they remain forever concealed from us."

Although scholars had been studying the heavens for centuries, no one had discovered that Mars's orbit was not a circle, but ancient and medieval astronomers cannot be accused of ignoring data to adhere to an erroneous assumption. The amount by which a planet's orbit, even the orbit of Mars, departs from a circle is extremely small. The errors in the observations available prior to Tycho were at least as great as ten minutes of arc, and this margin of error—of which astronomers were well aware—made it impossible to discern that the orbit was other than circular. Tycho's observations were trustworthy to within two or three minutes of arc, a great improvement on the ten-minute error tolerated before, but the discovery that Mars's orbit was elliptical would not come—as one might naively suspect—from Kepler's simply plotting a number of points where Tycho had found Mars and then failing in the attempt to draw a circle whose rim would pass through all of them. To discover the true orbit of Mars from Tycho's observations required a level of subtlety, insight, and inventiveness from Kepler that arguably has not been surpassed in the history of science.

The attack on the problem of Mars as it was originally assigned by Tycho to Longomontanus and Kepler was not intended to address the question whether or not Mars's orbit was circular. It involved two basic calculations. The first was the position of Mars's apsidal line— the straight line passing through its aphelion, equant, eccentric, Sun (for in the Tychonic system Mars orbited the Sun), and perihelion. The second was the extent of Mars's eccentricity (how far the center of Mars's orbit was from the Sun). If Ptolemy was right about the position of the equant in relation to the center of the system and the eccentric, they could expect to find that Mars's equant was twice as far from the Sun as the eccentric was.

From the time Kepler first began working with Tycho's data at Benatky, he chose to let the tight constraints of mathematical/geometrical logic and precise observations be his primary guides and to give them, for a while, precedence over the ideals of symmetry and harmony. However, he had by no means abandoned those ideals. He would continue to measure his theories against them and be uncomfortable when his results did not survive the test. Kepler was setting a precedent still followed in science, where symmetry, harmony, and logical beauty are not the most important criteria for judging whether a theory is correct, but where there is suspicion if those hallmarks are absent. Kepler had also not given up on his former theories: He hoped with Tycho's observations to be able to find out whether his polyhedral and harmonic theories were correct.

In other ways as well, Kepler did not begin his assault on the orbit of Mars with his mind a tabula rasa. Though he described his efforts as a voyage of exploration, he, like most explorers, knew the direction he thought he was heading, if not exactly what he would find there. He had already come to believe that understanding planetary motion required knowing the physical explanation for the motion, and he had already reached the conclusion that an essential part of the physical explanation was a force residing in the Sun that caused the planets to move in their orbits.

Kepler did not think that mathematical rigor, ideals of symmetry and harmony, and the search for a physical explanation were incompatible. He dove into a body of data that he trusted implicitly, though it was not his own, placing his bets that if his math was good enough and his instincts correct, he would come out the other side with his convictions about a physical explanation confirmed, and also clutching the trophies of symmetry and harmony. No one had traveled this particular route before. Kepler was not merely using science to find answers; he was working out what "science" was and would be, for himself and future generations.

Unlike Galileo when he wrote his famous *Dialogo,* Kepler did not intend the book he first referred to as "Commentaries on the Theory of Mars," which later evolved into *Astronomia Nova,* for the popular market. His target audience were his fellow early-seventeenth-century astronomers, men well versed in mathematics and planetary astronomy, including Copernicus's astronomy, although most of them did not think Copernicus had been suggesting anything "real" when he put the Sun in the center. Kepler realized that a battle to discredit the ancient models of astronomy had best take place on familiar ground, with familiar weapons, and not look like a battle. Hence his book would have to spend some time meandering benignly through intellectual landscapes where his contemporaries felt comfortable, not to mention where they were capable of recognizing and coming to trust his own knowledge and skill.

Kepler's mind was well suited to this kind of discourse. At age twenty-six he had written, "There was nothing I could state that I could not also contradict." He had no problem setting more traditional models in a fair and serious manner against Tycho's data as a way of discovering for himself and leading his readers to see that no astronomer, even with the utmost mathematical and geometrical maneuvering, could rest his case with these theories. He hoped to convince his readers that a theory had to be able to survive the common-sense questions of what might actually be going on in the heavens among these huge, real bodies . . . and *why.* Thus Kepler thought he would set the stage for his new astronomy, and it would prove to be accurate to the limits of Tycho's observations. At the outset, a confident Kepler who had promised to finish this book by Easter of 1602 had no way of knowing how long it would take him, how doubtful its outcome, how much ingenuity it would require of him, how many times he would fail, and how new his new astronomy would have to be.

To understand Kepler's achievement in writing *Astronomia Nova*, one must bear in mind that for most of Kepler's contemporaries there was no reason to wonder what the orbit of Earth was like. In the Ptolemaic system and even in the Tychonic system, Earth had no orbit. It sat still. Copernicus had implied that Earth orbited the Sun like the other planets, but he had not carried through with this in his mathematical analysis. It was a matter of extreme interest to Kepler whether or not Earth did indeed orbit *like the other planets,* speeding up when closer to the Sun (at perihelion) and slowing down when farther away (at aphelion). He believed it must and that if he used the true, visible Sun as his reference point he would find he was right. It had been a significant move when, at Benatky in the spring of 1600, he asked for and obtained Tycho's permission to use the true Sun when calculating Mars's orbit. He again used the true Sun when he determined that Earth does move more quickly when it is closer to the Sun and more slowly when it is farther away. Earth, at least in this regard, is nothing unique. It is just a planet. Kepler would spend many pages persuading his readers that it was better to use the true Sun. A physical explanation demanded it.

Such an approach also made it ridiculous to use devices consisting of invisible circles centered on invisible points on invisible circles on invisible points. However, it was not only to keep his readers with him that Kepler continued to use traditional Ptolemaic tools such as epicycles and equants, for Kepler needed all the mathematical and geometrical help they could provide to find his way to an astronomy that could do without them.

While Kepler had a superb mathematical mind, and improved his skills continually as he wrote *Astronomia Nova,* much of modern mathematics, including calculus, had not been invented yet. He also lacked the concept of inertia, though his contemporary Galileo understood it. Kepler did not visualize a universe in which an object keeps moving in a straight line at the same speed unless something comes along to affect that movement. Instead he thought an object sat still unless something moved it, and if that something ceased to move it, it reverted to stillness. In looking for the causes of motion he had to ask not only, "Why does it move as it does?" but also, "Why does it move at all, and why does it keep moving?"

Perhaps most significant of all, as Kepler set out on his quest, he was, as he put it, "armed with incredulity"—and hence ready to question all assumptions from the past as well as the theories and discoveries he made himself along the way.

Kepler's work with the Mars data had begun at Benatky in the unhappy spring of 1600. If the hand of God was in this enterprise, as Kepler believed—and Tycho Brahe would not have disagreed—then God's work on it had begun considerably earlier, and *Astronomia Nova,* which would be Kepler's masterpiece, was the result toward which divine purpose had been moving Tycho and Kepler with the most intricate and unlikely maneuvering for at least fifty years.

No one could possibly have obeyed Tycho's dying plea—"Let me not seem to have lived in vain"—more magnificently than Kepler did, but as Tycho had feared, he did it his own way, not the way that Tycho had intended, and history would celebrate the Copernican revolution, not the Tychonic revolution.

20

ASTRONOMIA NOVA

1600–1605

AFTER BEGINNING *Astronomia Nova* with a strong demonstration of how important the role of the Sun is, something he intended to hammer home even more emphatically later, Kepler turned his and his readers' attention immediately to Ptolemy. His strategy was to improve and generalize Ptolemy's theories as Ptolemy himself might have done had he had Tycho's data—an exercise of which no Ptolemaic astronomer could disapprove. Kepler appropriately titled this section of his book "In Imitation of the Ancients."

Kepler felt obliged to preface his "imitation" with six chapters describing and justifying the rigorous reworking he had given Tycho's observations in order to use them effectively. One problem was the necessity of undoing some choices and corrections made by Tycho and his assistants. For example, Kepler made extensive use of observations of Mars at opposition. Opposition is normally loosely defined as being when a planet is on the opposite side of Earth from the Sun. However, at opposition, Mars is rarely *directly* opposite the Sun, because of its latitude north or south of the ecliptic (review figure 7.6). Tycho and his assistants knew they had to compensate for this,

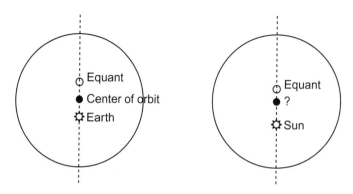

Figure 20.1: Ptolemaic astronomers placed the center of a planet's eccentric orbit precisely halfway between its equant and the Earth (left). Kepler wanted to find out for himself where the center of Mars's orbit lay in relation to its equant and the Sun (right).

but they had failed to do so consistently, nor was it always clear in the logs whether they had or had not.

In imitating the ancients, Kepler began by pointedly choosing not to imitate them in one important detail: Ptolemaic astronomers placed the center of a planet's circular eccentric orbit precisely halfway between the equant point and Earth. Kepler wished to make no such assumption but to discover for himself where Mars's orbital center lay. His work on this model began when Tycho was still alive, and it continued in the months after Tycho's death. Kepler was still thinking in terms of a circular orbit.

To develop his model, Kepler chose observations of Mars that Tycho had made when Mars was at opposition. At this time Mars, Earth, and the Sun were in line, and an earthly astronomer saw Mars in approximately the position it would appear at that moment were he or she standing on the Sun. There were ten oppositions of Mars in Tycho's log, the first in 1580. Hoping to measure Mars's parallax, he had made observations with extreme care. Kepler would later make two more himself. Determining the position and time an opposition occurred, with the precision Kepler required, was no easy matter. It was a considerable achievement, requiring a great deal of skill and understanding, to deduce this information, for it could not be found directly. No astronomer was able to see Mars and the Sun at the same time during opposition, when the two bodies are on opposite sides of Earth, nor was it possible to see the background zodiac stars behind the Sun, for the Sun is too bright.

Though the calculations involved in developing Kepler's new model were long and difficult, the family budget did not allow him to hire a permanent assistant to share the mathematical drudgery. "If you are wearied by this tedious procedure," Kepler begged his readers, "take pity on me who carried out at least seventy trials." Finally he was able to make a simple model that agreed with four of Tycho's observations of Mars at opposition. From this model he could calculate, for any given time, where Mars would be seen from the Sun—its "heliocentric longitudes." Kepler checked the theory against the remaining six observations from Tycho, and later against the two of his own, and he found his model agreed within the limits of those observations.

However, Kepler declared the model unsatisfactory. It was true that if one were standing on the Sun, one would see Mars in the positions predicted by his theory. But there was more to finding the correct location of a planet than knowing its position against the background stars, as seen from the Sun. Kepler wanted to know how *far* the planet was from the Sun at these positions.

In answering this question, Kepler discovered a serious discrepancy. His new theory indicated that the center of Mars's orbit was six-tenths of the way from the Sun to the equant point rather than halfway between, as Ptolemaic astronomers assumed. However, to obtain the correct *distances,* he had no choice but to put the center of the orbit right back where Ptolemaic astronomy had traditionally put it. And when he did *that,* his theory no longer predicted correct heliocentric longitudes (positions of the planet as seen from the

Sun). The errors were as large as eight minutes of arc.* Kepler's faith in Tycho's observations did not allow him to let this pass.

The failure was not a defeat. In fact, Kepler had his readers where he wanted them, forced to admit that something new was required: "After divine goodness had given us, in Tycho Brahe, so careful an observer that from his observations the error of calculation amounting to eight minutes betrayed itself, it is appropriate that we recognize and utilize in a thankful manner this good deed of God's—that is, we should take pains to search out at last the true form of the heavenly motions." Kepler's model, which he dubbed his "Vicarious Model" or "Vicarious Hypothesis," had brought him to a crossroads.

Kepler then told his readers that a "renovation of the whole of astronomy" must begin at home. Suppose Tycho's observations—indeed all observations—were (as Copernicus had suggested) made from a moving Earth. It behooved astronomers to be sure that their picture of this motion was correct. A flaw in that understanding would cause errors in any other astronomical work. In that interest, Kepler now changed directions and asked his readers to look toward Earth as though they were standing on Mars. With a brilliantly conceived triangulation from Earth's orbit to Mars,† Kepler demonstrated that Earth's orbit and motion were like that of the other planets. Though Ptolemy, Copernicus, and Tycho had all thought that the center of Earth's orbit was the same as its equant point, Kepler's results showed that it lay instead somewhere in the middle between the equant point and the Sun, and that was where astronomers had traditionally put the center of the orbit of a *planet*. Even more significantly,

---

*Eight minutes of arc is the equivalent of a little less than the thickness of a penny held at arm's length and viewed edgewise.
†In order to appreciate fully the manner in which Kepler used Tycho's observations, it would be necessary to follow him in far greater detail than is possible here. Appendix 3 describes in a much simplified fashion just this one short phase of the work to give a flavor of the interplay between Tycho's data and Kepler's use of it. The process did not "draw Earth's orbit" for Kepler, though it might seem it could have. It did not lead him to conclude that the orbit is elliptical. That would come later.

Kepler had found that Earth was speeding up when it came closer to the Sun and slowing down as it moved away. In other words, it was *moving* like a planet. The discovery that Earth behaves like a planet was a truly momentous advance and a strong argument on behalf of Copernican astronomy. In his book, Kepler had cleverly introduced his readers to that argument before they had a suspicion of where they were headed.

Kepler saw that the speeding up and slowing down had a predictable mathematical regularity to it. The speed of Earth at aphelion and perihelion was inversely proportional to its distance from the Sun. He decided that this rule surely had to apply not only at aphelion and perihelion but to the entire orbit. Kepler had arrived at his so-called distance rule.

Whether or not this tentative "rule" would turn out to be correct, Kepler had clearly, in getting there, become a virtuoso in the use of Tycho's observations, devising ingenious ways to exploit their unique accuracy and comprehensiveness, bringing together sets of observations so that the whole amounted to much more than the sum of the parts, honing his mathematical skills and his creativity against the constraints of this precise data. Such mastery of the creative nexus between observation and theory has seldom been achieved and never surpassed in all the history of science. Tycho, had he been alive and able to see beyond his bias for the Tychonic system, might well have cheered for joy, for Kepler was asking questions that no one had thought to ask before, and still the observations required to answer them were right there in Tycho's log.

Kepler did not turn directly to the question of whether his distance rule was correct, for he was determined to prise open a door into a new era of science where the search for physical explanations was of paramount importance. He was convinced that the true motion of the planets would elude him and all other astronomers until they knew the answer to the question of what was *causing* that motion. So he chose at this juncture to shift his focus to the search for physical explanations, thinking of this as by far the most urgent part of his work.

Kepler was wrong to believe that understanding the physical explanation for planetary motion had to come before knowledge of what that motion was. His discoveries of his three laws of planetary motion would precede by about three-quarters of a century Newton's discovery of the physics that lay behind them. However, it seems that Kepler's *attempts* to discover and understand the physical explanations, though often futile, were a necessary step in the process through which he discovered his laws.

One immediate result of Kepler's thinking along physical lines was that it pointed up how ridiculous and *un*physical previous descriptions of planetary motion were. He reasoned that since changes in a planet's distance from the Sun appeared to dictate changes in its speed, the cause of the motion must be in one of the two bodies. Though he had already made up his own mind on that score, in his book Kepler paused to consider that idea in the contexts of the different planetary models. In the Ptolemaic system, for example, if the force that moved a planet in a circle resided in a body at the center of the circle, then it was difficult to conceive how a planet could possibly move in a circle with no body at its center—an epicycle, for instance. It was even worse if the planet must change its speed as it circled in the epicycle. It had clearly become impossible to take Ptolemy seriously, and Kepler sent that ancient genius bumping off on a trick unicycle with the wheel attached off-center to nothing at all. "The Sun will melt all this Ptolemaic machinery like butter," wrote Kepler, "and the followers of Ptolemy will disperse, partly into Copernicus's camp, partly into the camp of Brahe." Kepler carted the epicycle off the stage, but he didn't throw it away.

The Tychonic system fared hardly better. The idea that a planet-moving force residing in the Sun caused the planets to orbit worked fine for the five planets that Tycho's model had orbiting the Sun, but when the Tychonic model required the Sun, in turn, to orbit around a stationary Earth, the arrangement floundered unless there was a separate force in the Earth to move the Sun, a force that did not affect the other planets. The Tychonic system was geometrically equivalent to the Copernican system, but Kepler saw that it was no match in terms of the possibility of a physical explanation. It was simply, distressingly, unlikely. Kepler might have imagined Tycho whispering urgently in his ear that the Moon circles Earth, not the Sun, and this posed a parallel problem. But Kepler put that off until another time and another book.

Kepler considered what the planet-moving force might be. It had to emanate through space in the way he had discovered light does. The strength of gravity, like the brightness of light, does fall off as the inverse square of distance, but Kepler did not discover that it does, and hence failed to arrive at the modern concept of gravity, though he came so close as to state, "If one would place a stone behind the Earth and would assume that both are free from any other motion, then not only would the stone hurry to the Earth but the Earth would hurry to the stone; they would divide the space lying between in inverse proportion to their weights." In spite of his comparison with light, he decided that the strength of the force felt by the planets fell off as the simple inverse of distance.

Kepler felt obliged to justify that conclusion for the obvious reason that when he studied the relative speeds of the planets, and the relationship for a single planet between its speeds in various parts of its orbit, he found that the planets' speeds *did* reflect a simple inverse relationship to distance from the Sun. There was no empirical evidence for an inverse square law. Gravity works in a more indirect way, so that the inverse square law, though it is correct, shows up only indirectly in planetary distances and speeds.

Kepler speculated that the Sun must rotate. He asked his readers to think of a lecturer surrounded on all sides by an audience. Those in the direction he is facing "see his eyes" while others "lack the aspect of his eyes." If he turns, his head turns, and his gaze sweeps the crowd.

Figure 20.2: If the strength of the force fell off as the inverse *square* of distance, planet B, twice as far away from the Sun as planet A, would feel only a fourth as much of the force as planet A.

Likewise, wrote Kepler, the rotation of the Sun caused the force that moved the planets to sweep around. The planets could not be rigidly tied to the Sun by this force, as though fixed at the ends of spokes of a wheel, because they moved at different speeds. The Sun therefore had to be rotating at a speed that got ahead of them. Because the planets were "prone to rest" (recall Kepler's lack of the concept of inertia), they lagged behind.

A rotating body that exerts a force at a distance through empty space, affecting closer things more strongly and giving out its force in the shape of a sphere, reminded Kepler of reading he had done recently about magnets, in books by Jean Taisner and the Englishman William Gilbert. Kepler never decided that the planet-moving force was magnetism, but it was encouraging that such a force was known to exist and that Gilbert had recently shown that Earth was a magnet. It was reasonable to think that the Sun could exert a similar force. Kepler's suggestion was that as the Sun rotated, a field of magneticlike emanations coming from it also rotated, in turn moving the planets.

Having stated that the force propagates in all directions, not just along the ecliptic, Kepler was obliged to account for the fact that the planets are not spread out in all directions from the sphere of the Sun but instead all orbit near the plane of the solar equator. Kepler's answer, using an analogy in which Earth replaced the Sun, was that if a planet were too far "north" or "south" it would be affected by the motion on the other side of the rotating globe. It would feel conflicting directions in the force that reached it, which over the "poles" would result in complete confusion. Hence Kepler felt that the planets could not help but end up orbiting only near the plane of the Sun's equator, where they were not affected by the opposite stream of motion on the other side of the Sun.

Kepler was still left with a stubborn problem. He had an explanation for why a planet more distant from the Sun would move more slowly than one closer, and why a planet's speed would vary as its distance from the Sun changed. But he lacked an explanation for why a planet's distance from the Sun *should ever change at all.* If the only thing at work were a planet-moving force residing in the Sun, then the planets would be carried around in circles centered on the Sun, never speeding up or slowing down. Something as yet unknown was alternately moving the planets toward and away from the Sun, to distances where the Sun's planet-moving force was stronger or weaker. He speculated that a planet might have a mind, or some other non-mental mechanism, yet if this were so, there was still a question how that mind or mechanism would know how distant it was from the Sun. Yet a planet seemed mysteriously able to decide how far it should move away before coming back, and when it was close enough to move away again. Kepler could think of only one way in which distance from the Sun shows up clearly, and that is in the apparent size of the Sun's disk. He conceded that there might be ways of "perceiving" that humans knew nothing about.

An explanation for planetary movement that endows a planet with a "mind" to help steer it through the heavens does not sound to modern ears like a "physical explanation." Kepler was not fond of the idea himself, but he could not dismiss it.

At this point in his book, Kepler decided that he had, for the moment, done all he could in the way of pursuing physical explanations and needed to return to the problem of describing the planets' true motions mathematically and geometrically.

How much time it took a planet to travel a given distance along its orbit depended on how far it was from the Sun, so much was clear. The planet was slightly changing its distance from the Sun, and hence its speed, continually all the way around the orbit. To get a handle on these changes, Kepler needed integral calculus, but that would not be invented for at least three-quarters of a century. He nonetheless found another way.

A circle has 360 degrees, so Kepler divided the circumference of a circular orbit into 360 equal arcs. Then he laboriously proceeded to calculate the distance from the Sun (off-center in that circle) to each of these separate arcs, as if measuring the length of every spoke of a wheel that has an off-center hub. Of course he had to calculate only 180 of the 360 "spokes," since planet-Sun distances on the other side of the orbit would be the same. Having done that, he could, for instance, imagine a planet starting at aphelion and passing through the first 30 of the 360 degrees of its orbit. The sum of those 30 orbit-to-Sun distances was to the sum of all 360 distances, as the time it took the planet to move those 30 degrees was to the time it took the planet to complete an entire orbit of 360 degrees. Even for Kepler this procedure became unendurably tedious and complicated. He decided to look for a shortcut.

Kepler recalled a method that the ancient mathematician Archimedes had used to calculate the area of a circle. Archimedes reasoned that a circle was made up of an infinite number of isosceles triangles* with their bases on the rim of the circle and their apexes at the center. Knowing he must use something more manageable than an infinite number, Archimedes, like someone who has baked a pie for a few that many show up to eat, divided the circle instead into very fine, equal isosceles triangles, again with their bases on the rim of the circle and their peaks at the center.

It went without saying that when Archimedes combined a few adjacent triangles of equal size, he was doing two things. One, he was combining their bases along the rim of the circle. The more triangles

---

*An isosceles triangle is a triangle having two sides of equal length. (See figure 20.3[a], though it does not show an infinite number of triangles.)

he combined, the longer the portion of the rim taken up by those bases. If he combined all of the triangles, he would have taken up the complete circle. Two, he was combining the areas of the triangles. If he doubled the number of triangles in the group, the area doubled; triple the number, and the area tripled; and so forth. Clearly there was a relationship between the amount of the rim that the combined bases covered and the area of the combined triangles. For example, for an arc twice as long, the area would be twice as big.

Kepler suggested that combining the triangles does more than combine their areas and combine the lengths their bases take up on the rim of the circle. He thought of the circle as a spoked wheel with an *infinite number* of spokes representing center-to-rim distances. No matter what size one made the triangles, *each* triangle would contain an infinite number of these spokes. Kepler, like Archimedes, could not compute with infinite numbers. However, though it was not possible to say how many spokes were contained in any one triangle or combination of triangles, it was reasonable to conclude that the more triangles you combined, the more of these spokes there were in the combined area. Even more precise than that: Combine two triangles of equal area, and the number of spokes doubles, and so forth. With this idea, Kepler had found a way to think about the relationship between orbit-to-Sun distances, the time that passed as the planet moved along its orbit, and areas within the circle. Whether this line of reasoning could be applied to an off-center orbit, where the triangles were no longer isosceles triangles, was problematic. And of course the whole point of the exercise was to illuminate the workings of an off-center orbit.

Kepler persevered, and he reached a tentative conclusion: A straight line drawn from a planet to the Sun, as the planet orbits, would sweep out equal areas of the circle in equal times. When he tried this rule with Earth's orbit, it worked. Though he was not yet nearly confident enough about it to declare that it was correct, and never even clearly stated it in *Astronomia Nova,* Kepler had arrived at

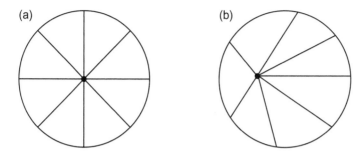

Figure 20.3: If the Sun were in the precise center of the orbit (a), the triangles would be isosceles triangles. But with the Sun off-center (b), the triangles Kepler was considering were no longer isosceles triangles.

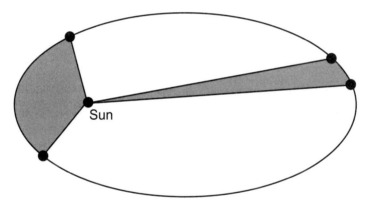

Figure 20.4: Kepler's area rule, shown here in its final form, as his second law of planetary motion, with an elliptical orbit. When he first arrived at the area rule, he was still trying to apply it to a circular orbit.

Imagine the planet moving around the orbit with a straight line drawn from it to the Sun. As the planet moves, so does the straight line. Watch the line move for, let us say, two minutes, then measure the area of the pie wedge it has "swept out." For every two-minute interval, the wedge will have that same *area,* but it will not always be the same shape, nor will the edge of it that touches the orbit always be the same length. Near the Sun the wedge will be fat and cover a long portion of the orbital path. Far from the Sun it will be thin and cover a much shorter portion, meaning that far from the Sun the planet is moving a shorter distance in the two-minute interval.

what has come down through history as his "area rule," his "second law of planetary motion." Confusingly, he did not discover his "first law" until somewhat later.

Kepler realized immediately that his old distance rule and this new area rule were not necessarily the same. The limits of observational accuracy made it impossible to judge which was correct for Earth. He knew he must look at the orbit of another planet.

At this critical moment in October 1602, Kepler was rudely interrupted by Tengnagel's return to Prague and his conclusion that Kepler had made no progress in the use of Tycho's observations. It is not difficult to understand why Kepler secretly kept the Mars data, judging, correctly, that Tengnagel would not notice at least for a while.

The old "problem of Mars" now offered a splendid opportunity. Because Mars's orbit was farther from being centered on the Sun than Earth's orbit was, a flaw in the area rule was more likely to reveal itself. Kepler used the area rule to compute where Mars should be in its orbit at given times during the 687 days the planet takes to complete the orbit, and then he checked these predictions against the heliocentric longitudes of his Vicarious Hypothesis. He found agreement when it came to certain parts of the orbit but not others. In fact, he was back to an eight-minute discrepancy! Again, it had come to a showdown: Either the circular orbit was wrong, or the area rule was wrong. Kepler could not rule out even the possibility that *both* might be wrong.

Though still far from completely trusting his area rule, Kepler decided to take the plunge and try a noncircular orbit. A triangulation like the one he had used earlier for Earth's orbit indicated that Mars's path was indeed not a circle but bowed in at the sides. Mars was like a racer who cheats by coming within the circle of a circular racetrack. Doing that while still having to make it around two goalposts (aphelion and perihelion) would change a circular race into an oval race.

The precise amount by which Mars was "cheating" in this race was fiendishly difficult to establish. Circles, except for size, are identical. "Oval," on the other hand, is a much less precise term. An ellipse is one kind of oval, the best defined geometrically and the one a man obsessed with harmony and symmetry in nature might be expected to assume was correct. Kepler did not. Not only did movement in an elliptical orbit appear to defy a physical explanation, but it also seemed too easy an answer. Kepler wrote to his friend David Fabricius that surely if the orbit were a perfect ellipse the problem he had been struggling with would have been solved long ago by Archimedes or Apollonius.

At the time he wrote that letter, in July 1603, Kepler had been forced to abandon the struggle with Mars, because Tengnagel had finally noticed that the Mars observations were missing and confiscated them. Kepler was working on *Astronomiae Pars Optica* instead. It was not until a year later that he had the Mars data again.

As Kepler resumed juggling ovals, which he would continue to do for the rest of 1604 and in the beginning of 1605, his frustration grew intense. His math was inadequate. He was suspicious of his area rule. He even had some doubts whether Mars's orbit made sense mathematically *at all*. An attempt he made to calculate Mars's positions degree by degree gave him unsatisfactory results and was the sort of procedure he despised. This was not geometry, and Kepler took issue with God on the matter, in words he might have used to comment about a human colleague: "Heretofore we have not found such an ungeometrical conception in his other works." Kepler had not changed in his intolerance of procedures or results that insulted his geometrical sensibilities.

Kepler resorted to working with an ellipse that he called the "approximating ellipse," to see what he might learn from the exercise. That presented a new problem. He had earlier (as he described it) been obliged to "squeeze in" his circular orbit as though he were holding a "fat-bellied sausage" in his hand and squeezing it in the middle so that the meat was forced out into the ends. With his approximating ellipse, he had squeezed the sausage too much. The correct orbit had to be something in between.

Kepler's desire for a physical explanation made his efforts more difficult. He had begun to think that the force resembling a magnetic force might account not only for the motion of the planets around the Sun but also for their motion toward and away from it. That was out of the question with an elliptical orbit. One of Kepler's uses of an epicycle as a computational device had led him to have rather high hopes for the magnetic hypothesis, so he brought yet another epicycle out of storage. That resulted in a "puffy-cheeked" orbit *(via buccosa)*. It is one of the ironies of scientific history that it was an error in his calculations that caused Kepler to reject this orbit. Kepler had reached chapter 58 when he wrote, "I was almost driven to madness considering and calculating this matter. I could not find out why the planet would rather go on an elliptical orbit!"

And then, "As if I were roused from a dream and saw a new light," a torrent of answers fell into place. An elliptical orbit halfway between his approximating ellipse and a circle had a feature that was deeply satisfying to one who loved geometric harmony. The Sun was one of its foci. Kepler had arrived at his "first law of planetary motion."

This, as Kepler put it, was "the sort of thing nature does." With this ellipse, the orbit made physical sense, supporting his conviction that a force residing in the Sun moves the planets. What was more, if the area rule was correct, this model agreed "to the nail" with the long-trusted heliocentric longitudes of his Vicarious Hypothesis. This one shape of orbit, and *only* this shape, got the planet to the right place at the right time. The man who had said of himself, "There was nothing I could state that I could not also contradict," had discovered a piece of incontrovertible truth.

At Easter 1605, the second Easter after the one for which he had promised his book, Kepler decided definitely on the ellipse. He finished the manuscript that year, adding a subtitle to emphasize that his "New Astronomy" was "Based on Causes, or Celestial Physics." Kepler ended *Astronomia Nova* with the hope that God, having so richly endowed his creatures with analytical brains and insatiable curiosity, and

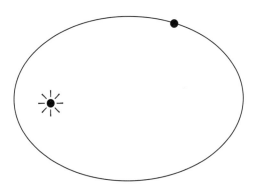

Figure 20.5: Kepler's first law of planetary motion: A planet moves in an elliptical orbit, and the Sun is at one focus of the ellipse.

his Creation with surpassing beauty and ingenuity, would allow humans sufficient time on this Earth to resolve questions he, Kepler, had not yet been able to answer.

The adventure begun on Hven when Tycho first made the decision to train his fabulous instruments on Mars had taken Kepler to a new astronomy. He had made sense of the positions of Mars spread over many pages of observational logs. In this miraculous cohesion of observations and mathematical theory, the numbers, in the words of Kepler scholar Max Caspar, "no longer stand together unrelated but rather each can be calculated from the other." The limits of accuracy of Tycho's observations had turned out to be exactly right for the task Kepler undertook: "[They were] narrow enough so that Kepler could not afford to neglect those very important eight minutes . . . but had they been considerably narrower, he would certainly have been caught in a fine meshed net, because in many of his calculations he would no longer have been permitted to overlook certain inaccuracies, as was necessary for the progress of his research." Nevertheless, the precision of Tycho's observations made it possible for Kepler to find the elliptical orbit of Mars even though it is so near to being a circle that any

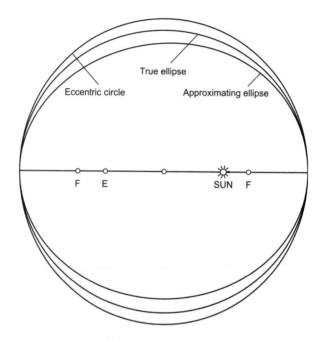

Figure 20.6: Comparing the eccentric circle, the true ellipse, and the approximating ellipse. F designates the two foci of the approximating ellipse. E and the Sun are the foci of the true ellipse. The drawing greatly exaggerates the eccentricity of the ellipses. Correctly drawn at this scale, they would be impossible to distinguish from one another or from a circle.

drawing of it on the page (such as the ones in this chapter) that makes it look even slightly elliptical is a gross exaggeration.

Kepler's discoveries that Earth behaves like a planet, and of his first and second laws of planetary motion, were towering landmarks in human intellectual and scientific history. He had indeed plumbed the depths of a complicated universe and found harmony. He had also given Tycho Brahe his earthly immortality.

21

THE WHEEL OF FORTUNE
CREAKS AROUND

1606—1618

THE AGREEMENT STRUCK earlier between Kepler and Tengnagel required Kepler to submit the manuscript of *Astronomia Nova* to Tengnagel for approval. Tengnagel was not pleased: The book clearly argued for the Copernican rather than the Tychonic system. The difficulty was settled when Kepler agreed to allow Tengnagel to write a preface. Hence *Astronomia Nova*, like Copernicus's *De Revolutionibus*, begins with an unpromising warning, in this case that readers should "not be swayed by anything of Kepler's, especially his liberty in disagreeing with Brahe in physical arguments." Both Tengnagel and Osiander (in the case of *De Revolutionibus*) emerge looking foolish in prefaces to two of the most significant astronomy books in history.

The publication of Kepler's book moved at a snail's pace. The printing didn't even begin until 1608. It was Rudolph's right to distribute all copies of a book by his imperial mathematician, but when it appeared in the summer of 1609, Kepler had to give the entire edition back to the printer in Heidelberg to sell to cover unpaid costs.

January 1610 marked ten years since Kepler had first arrived in Prague in Hoffmann's carriage. There was abundant reason to celebrate the anniversary. Kepler's reputation as scientific heir to Tycho Brahe, and the books Kepler had written, had elevated him from the status of an impoverished provincial mathematics teacher to that of a celebrated figure in educated circles all over Europe and in Britain. This success redounded to the emperor's credit as well. Rudolph lavished praise on Kepler and granted him a bonus of two thousand talers, which would have been splendid had it been paid.

Sadly, Kepler's decade of superb scholarly achievement, warm friendships, and almost universal respect was marred not only by his problems collecting his salary, but also by the decline of both his wife and his patron. Barbara had made some friends among the more pious Protestant women of Prague, but after a decade in the city she was still miserably homesick. In spite of her discontent, the Keplers had seldom been out of Prague at all during the years since Tycho's death, except for a sojourn in Moravia when the plague came back in 1606. In the autumn of 1607 they had moved to new lodgings near the great bridge over the river. Ludwig, their second son, was born there that December. In 1608 Regina, Barbara's daughter and Kepler's stepdaughter, married Philip Ehem, who was descended from a prominent Augsburg family and was currently a representative at the imperial court for Elector Frederick IV of the Palatinate. There were many reasons for happiness and pride, yet Barbara suffered chronic bad health and deepening depression.

As for the emperor, by the time he granted Kepler's celebratory bonus, Rudolph had been stripped of almost all his power. Oddly enough, his habitual state of political indecision and inaction had stood him in rather good stead for many years. He had kept up an endless, stalemated war with the Ottoman Turks and held together, often only by failure to act or react, an empire forever threatening to disintegrate. Rudolph had always been quirky and pathologically shy, but over time his indecision and stubbornness had degenerated to paralysis, and his mental state had continued to deteriorate until he was rumored to be insane. Though he was such a recluse that it was difficult to confirm or deny the rumor, he had clearly become a threat to the royal house of Hapsburg and the empire.

The Austrian Hapsburg family had met in secret in April 1606 and agreed to recognize Rudolph's younger brother Matthias as the head of the family instead of Rudolph. Two years later, Matthias led an army from Vienna into Bohemia to within a day's march of Prague. Rudolph abdicated, ceding to Matthias the kingdom of Hungary and the archduchies of Austria and Moravia, keeping only Bohemia (including Prague), Silesia, and Lusatia for himself, with Matthias named as his successor there.

In the spring of 1610, a few months after Kepler's anniversary celebration, nature for a time upstaged domestic or political concerns, at least for Kepler and some of his close acquaintances. On March 15 a carriage stopped at Kepler's door. Its passenger was Kepler's friend Wackher von Wackenfels, who was an imperial councillor twenty years Kepler's senior, a distant relation, and a brilliant man with many scholarly and scientific interests. Von Wackenfels stuck his head out of the carriage window to call Kepler to come down immediately, for there was stupendous news: Galileo Galilei, with his new telescope, had discovered four new planets. Von Wackenfels and Kepler were so overcome they could do little more than babble with excitement. Kepler's enthusiasm was tinged with anxiety over whether these discoveries were planets, or moons around one of the other planets. Von Wackenfels was not sure, but Kepler said they surely must be moons, because he had established with his polyhedral theory that there could be only six planets.

A copy of Galileo's book reporting the discovery—*Sidereus Nuncius* (The Starry Message)—reached the emperor, who loaned it to Kepler. Kepler's own copy arrived soon after via the Tuscan ambassador with a request that the imperial mathematician send his opinion back with the same courier before the week was out. To Kepler's relief, he read that the "planets" were four moons orbiting Jupiter.

Kepler's response enthusiastically agreed with Galileo that the

Galileo Galilei

discovery of Jupiter's moons supported Copernican astronomy, for it was evident that not everything in the universe was revolving around Earth. The fact that through a telescope the planets looked like disks but the stars remained points—no larger than they appeared to the naked eye—indicated that the stars were indeed as far away as Copernican theory required. As for Kepler's theories, it no longer seemed such an oddity that Earth should have its own version of a "planet-moving force" to keep a moon orbiting. Kepler suggested that Jupiter must have intelligent inhabitants; otherwise, why should God have given Jupiter moons that would go unappreciated by any creature except the few Earth dwellers who had telescopes? Kepler was not among those fortunate Earth dwellers.

Kepler's letter was ready on April 19, in time for the courier's return trip. Galileo's response was not so rapid in coming. It took him four months, but it was effusive: "I thank you because you were the first one, and practically the only one, to have complete faith in my assertions."

There was widespread curiosity about Kepler's reaction to Galileo's discovery. Kepler published the letter he had written to Galileo as a thirty-five-page book, *Dissertatio cum Nuncio Sidereo* (Conversation with the Starry Messenger). One of the most surprising things about it was that Kepler felt called on to defend the use of the telescope as a reliable scientific instrument. Some astronomers had voiced suspicions that Galileo's discoveries might be nothing more than artifacts of the instrument, not something really there. Kepler's reassurance about the use of telescopes, though he had seen only inferior models in Prague and was relying on his knowledge of optics, was almost as valuable as his defense of the plausibility of Galileo's discoveries. Kepler had hinted broadly to Galileo that he would like to have one of the telescopes Galileo was sending to important people all over Europe, but Galileo did not send him one, so, unable to confirm the discoveries himself, "plausibility" was as far as Kepler could go.

In the late summer and early autumn, soon after receiving Galileo's belated reply, Kepler was able to borrow a telescope Galileo had sent Elector Ernst of Cologne, duke of Bavaria. Kepler called together the mathematician Benjamin Ursinus and several other friends, and they viewed Jupiter from August 30 to September 9. To avoid being misled by one another, they agreed that each man would look through the telescope and, without comment, draw in chalk on a tablet what he had witnessed through the lenses. Only after everyone had done this were the drawings compared.

Kepler simultaneously followed up on ways two lenses could be combined to magnify images. That same August and September he wrote a book on the subject, *Dioptrice,* which was published in 1611. Kepler warned in the preface, "I offer you, friendly reader, a mathematical book, that is, a book that is not so easy to understand." His approach in *Dioptrice* was indeed rigorously mathematical. The book contained the first detailed optical theory of two lens systems and a new, improved telescope design, later known as the "astronomical" or "Keplerian" telescope.

Kepler made several attempts to continue the correspondence with Galileo, but Galileo, except for a short letter seventeen years later recommending a student, apparently never wrote Kepler again. His silence may have been partly a reaction to Kepler's first letter about the moons of Jupiter, for Kepler had not hesitated in a friendly fashion to set the historical and scientific context straight by mentioning other researchers whose imaginative thinking might have helped lead Galileo to his discoveries. Mästlin congratulated Kepler on having "pulled out Galileo's feathers exceedingly well," and since Galileo had chronic difficulties recognizing fine gradations in friendship and support, he may have seen Kepler's letter as threatening. Later, in an appendix to a small book defending Tycho Brahe's theories about comets, Kepler upbraided Galileo for erroneous ideas about them and pointed out that the phases of Venus that Galileo insisted were strong support for Copernicus's model were just as much in accord with Tycho's.

Galileo was not a man to take well to such reproofs, but he had had little support from the scientific community and had probably never met anyone who was his scientific equal. If Kepler and Galileo had exchanged more letters and ideas, Galileo might not have gone on for the rest of his life believing that planetary orbits were circles and that the Moon had nothing to do with the movement of the tides. And Kepler would have learned about inertia.

❦

AT NEW YEAR 1611, Kepler gave his friend von Wackenfels a whimsical gift, a letter about why snowflakes are hexagonal. Kepler's writing reflected his cheerful mood that early winter as he invented the puns that he put in his letter, anticipating Wackher's laughter when he read it. *Strena, a New Year's Gift; or, On the Six-Cornered Snowflake* was both a delightful bauble and a pioneering study in what would become the science of crystallography. Kepler could not know that the time spent on this letter would be the last happy hours, and *Strena* the last lighthearted achievement, of his years in Prague among such friends as von Wackenfels.

That winter, Rudolph made a desperate and foolish move. He plotted with his cousin Archduke Leopold V, bishop of Passua, to bring an army to Bohemia. What Rudolph hoped this would accomplish is not clear. In February, while the populace of Prague waited in dread as Leopold's undisciplined, unpaid soldiers ravaged the countryside and neared the city, tragedy struck the Keplers. Barbara was still recovering from Hungarian fever, which she had caught just before New Year, when the three children came down with smallpox. Eight-year-old Susanna and three-year-old Ludwig recovered, but six-year-old Friedrich, who had been a particular delight to Kepler, died on February 19. It had been many years since he and Barbara had lost their first two infant children. This time Barbara was less resilient. Kepler wrote that she was "wounded to the depths of her being by the death of the little boy who was half her heart to her." She slipped into even deeper depression.

Within days of Friedrich's death, Leopold's troops were in Prague and occupied the area surrounding the palace and the Lesser Town nearer the river. Bohemian Protestant vigilantes banded together in other parts of the city, including the area around the Keplers' house, ostensibly for defense purposes but also to loot cloisters and Catholic churches in the Old Town. The streets of Prague became bloody battlefields as the two groups fought for turf, while in the imperial palace the atmosphere was thick with madness and ruin. The emperor paid off Leopold's men (the treasury, for once, was responsive), and they departed, but Rudolph's reign was over. Preparations were made to crown Matthias king of Bohemia. Life as Kepler had known it in Prague, on both the personal and public levels, had come to an end.

In the spring of 1611, as Kepler put out urgent feelers in an attempt to make provision for his and his family's future, he once again looked hopefully to his native Württemberg. With a list of achievements and honors to match any in Europe, he hoped to be welcomed home with a professorship or a political appointment in the ducal court.

In April that door was emphatically slammed in his face, his application denied because of his earlier admission, still on record, that he believed a Calvinist also was a "brother in Christ." In Lutheran Württemberg this amounted to a criminal view. A man espousing it might spread his poisonous ideas among his students. Ironically, Kepler was excluded for following one of the basic tenets of Lutheranism, the principle of a "priesthood of all believers" in which it is every believer's right to interpret the Scriptures for himself.

Though there were other possibilities had Kepler had time and heart to pursue them, he accepted a teaching position in Linz in a school similar to the one where he had started his teaching career. It was not a university appointment of the sort one would have thought awaited a man of his stature, but he could look forward to being treated with respect, and the position of provincial mathematician was created especially for him. Since Linz was in Upper Austria, he could also retain the title of imperial mathematician. The new emperor Matthias could be expected to reconfirm that appointment. Kepler's contract called for him to "complete the astronomical tables in honor of the emperor and the worshipful Austrian House, [to benefit] the entire land and also for his own fame and praise." He was also charged with making a map of Upper Austria and "producing whatever other mathematical, philosophical, or historical studies were useful and suitable." As he made these arrangements, Kepler clung to the hope that life for Barbara in Linz would be more like it had been in her beloved Graz. For his sake, she had endured ten years in Prague, and it was her turn to find some measure of happiness.

In June, returning to Prague from a journey to settle matters in Linz, Kepler found Barbara again dangerously ill. Matthias's Austrian troops, now in the city to establish peace, had brought a contagious

A portrait believed to be of Johannes Kepler, painted by Hans von Aachen around 1612.

fever. Barbara had insisted on helping nurse the sick, and she had caught the fever herself. She died on July 3.

Kepler and what remained of his family did not make the move to Linz immediately, for Rudolph still needed him in Prague. To the last, Kepler remained loyal to his unfortunate patron, dividing his time between a grief-stricken home and a doom-stricken palace. With the political situation beyond hope, he nevertheless struggled to keep astrology as far as possible out of the heads of the gullible emperor and his closest advisers, and attempted to mislead Rudolph's enemies by informing them, counter to what he was actually finding in his astrology, that the stars still predicted long life for Rudolph and difficulties for Matthias. Rudolph died in January 1612. Matthias renewed Kepler's appointment as imperial mathematician but permitted him to leave Prague. In May 1612 Kepler, soon to be followed by his two children, aged eight and three, moved to Linz without the wife and mother for the sake of whose happiness Kepler had decided to go there.

Kepler's reputation as a Nonconformist with a tolerance for Calvinist views followed him to Linz. The Lutheran pastor asked him to recant on this issue and, when he refused, denied him the privilege of taking Communion. Though Kepler may have seemed to militant Lutherans to lean toward Calvinism, he was not a Calvinist. "It makes me heartsick," he wrote, "that the three big factions have so miserably torn up the truth among themselves that I have to gather the little scraps together wherever I find them." This devoutly Christian man, so at home with his God, found himself without a home in any earthly church and having to endure gossip in Linz about his religious plight and exclusion from Communion.

There was, however, a brighter side to his new situation. His salary of 400 florins was actually paid regularly. Also, in 1612, Kepler finally became custodian of all Tycho's observations. Tycho's son Georg was by then serving as the Brahe family's representative and proved to be a staunch ally at court when Jesuit astronomers with imperial support attempted to take Tycho's library, instruments, and manuscripts. Kepler, with Georg's help, kept the precious observational data.

The first task Kepler set his mind to in Linz was not scientific. With two young children, it was essential that he remarry. That concern occupied his thoughts for a year while he considered no fewer than eleven candidates. As a topic of conversation in Linz, the subject of Kepler's wooing eclipsed the subject of his religious problems. He weighed the advantages of each woman in his own mind and in letters to an acquaintance, discreetly referring to them not by name but by a number.

Number one was an experienced homemaker about his own age, but her breath stank. Number two was her daughter. She was too immature and accustomed to luxury. Numbers three and four were upstaged by number five, a serious, loving woman, whose humility, frugality, diligence, independent mind, and fondness for his children impressed him. However, her family was less respectable than number four and her dowry smaller. Kepler's friends told him he would be marrying beneath his station. So he favored three again, then four, but four had grown tired of waiting. Number six was immature and conceited, though there was a certain nobility about her. Back to number five. But then some friends suggested number seven, who was a noblewoman. When he failed to make up his mind immediately, she rejected him. Number eight was unsure whether she wanted to marry a man excluded from Communion. Number nine had a lung disease. Number ten was ugly and fat. Number eleven was offered and then the offer withdrawn because of her youth. Finally Kepler cast aside consideration of status, family opinion, dowry, and improvement of his social rank through marriage and declared that God had led him back to choose number five—Susanna Reuttinger.

The daughter of a cabinetmaker, Susanna was twenty-four years old, seventeen years younger than Kepler and only a year older than his stepdaughter Regina. She had been orphaned at an early age and lived most of her life as the ward of a baroness whose husband was one of Kepler's patrons in Linz. Kepler's stepdaughter Regina thought that Susanna was too young to be a good mother to his two young children, and there were some comments in Linz about the age difference, but Kepler had made up his mind. He loved Susanna, and he trusted her. They were married on October 30, 1613. The next summer Susanna gave birth to a daughter, Margarethe Regina, named after Kepler's stepdaughter.

Kepler's scholarly work slowed down during the time he wooed and chose among the eleven women, but he did not abandon it entirely. Once again, it was not the long-languishing Rudolfine Tables that he worked on primarily, nor was it the map that was in his contract. Instead, new inspiration came from an unlikely source. Traveling on the Danube, Kepler saw many differently shaped wine barrels on the riverbanks and became intrigued with the problem of how to express their volumes. Because of an unusually good wine harvest, he decided to install some wine casks at home and learned in the process that Austrian wine merchants measured only the diagonal length of a barrel, disregarding its shape. Kepler's book on the subject was not a huge popular success, and his superiors in Linz were not impressed that their mathematician was thinking about wine barrels rather than the Rudolfine Tables and the map. However, Kepler's study of the wine barrels did satisfy him that the old, simple way of measuring was adequate for Austrian wine casks. More significant for the future, the mathematics that Kepler developed in the process became an important step in the history of the development of integral calculus. Another unexpected side advantage was that when Kepler failed to find an interested publisher, he brought a printer, Johannes Planck, to Linz and published *Nova Stereometria Doliorum Vinariorum* (New Solid Geometry of Wine Barrels) himself.

When Kepler began producing calendars again in 1616, Planck printed those too. The production of the calendars was, according to Kepler, "a little more honorable than begging." It was his way of raising money for the next publishing effort he had in mind, for which there had been many requests, a textbook that would make the discoveries he had written about in *Astronomia Nova* more accessible to nonexpert readers, to "the low schoolbenches," as he put it. The first three volumes of the seven that would comprise *Epitome Astronomiae Copernicanae* (The Epitome of Copernican Astronomy) were ready for Planck in 1615, but the final pages of the seventh did not come off the press until 1621. By this time Kepler had added enormously to Copernican astronomy, and there was as much if not more of Kepler in the book than of Copernicus. *Epitome* was an influential book, read all over Europe.

Meanwhile, however, after completing the first three volumes of *Epitome,* Kepler turned to producing an immediate moneymaker, an

Ephemeris for 1618. An ephemeris consisted of tables giving the position of each planet for every day of the year, and such a book was an invaluable reference for both navigators and astrologers. Kepler knew that after the Rudolfine Tables were published, anyone who was not lazy would be able to calculate planetary positions without using an ephemeris, and sales on those would drop. If he wanted to profit from the sale of ephemerides (plural of ephemeris), now was the time. Printers normally did not have enough numbers on hand to print an ephemeris, and Kepler purchased his own set of numerical type for Planck to use. Kepler's confidence about producing these ephemerides year after year is evidence that he had made substantial progress studying the orbits of the other planets besides Mars and Earth.

❦

K E P L E R ' S   M A R R I A G E to Susanna began only a brief respite from personal problems. In December 1615, news arrived from Württemberg from his sister Margarethe, who was now Margarethe Binder: Their elderly mother had been accused of witchcraft.

Katharina Kepler's reputation as an unpleasant, meddlesome woman and her expertise in herbs and folk medicine had set her up as a target for the sort of grudges and gossip that in the sixteenth and seventeenth centuries in southern Germany could easily degenerate into a witch trial. Frau Kepler was probably an intelligent woman, but she was not a wise one, and she had no social skills. Those people with whom she associated, or who were willing to associate with her, were the dregs of society.

The crisis had begun when she had sided with her son, Kepler's brother Christoph, in a minor business dispute with one of her friends, Ursula Reinbold. Frau Reinbold, whom Kepler later dubbed "the crazy," had been imprisoned for a time for prostitution. One of the disadvantages of that profession was that she often had to abort illicit pregnancies, sometimes with the dubious help of her brother, a barber-surgeon, and at least once in the past with the help of a herbal mixture Katharina Kepler had provided. The present difficulty had erupted when Frau Reinbold, ill after a botched abortion that had nothing to do with Katharina, chose to believe that the potion Katharina had given her three and a half years earlier had been a "witch's drink" and was causing her present distress. She demanded that Katharina produce a "witch's antidote." Even though Frau Reinbold's brother held his sword at her throat, Katharina refused. To produce the potion would be admitting she practiced witchcraft.

After that frightening episode, in August 1615 Katharina, on the advice of Christoph and Margarethe's husband, a village pastor, took the wisest step available, though by no means a good one. She brought a libel suit against Ursula Reinbold.

In late December, when Kepler finally received Margarethe's letter informing him of these events, his "heart almost burst." He immediately wrote to the town senate of Leonberg, choosing his words skillfully to remind them they were dealing with a powerful and influential man and demanding they send him copies of all legal proceedings involving his mother. Several years before, he had revised his fanciful student essay about the Moon, using a plot device in which he described the narrator's mother as an old woman skilled in folk magic with the power to summon a demon. He agonized over the possibility that news of his essay, or even a copy, might have reached Württemberg or Leonberg.

As it happened, events in Leonberg were delayed for a time. The bailiff there, one Lutherus Einhorn, had been present when Frau Reinbold's brother held the sword to Katharina's throat. Not wanting to reveal his part in this affair, he managed to postpone the libel case until the following October.

Six days before the proceedings were finally to begin, Katharina was walking along a narrow path and met a group of girls who were carrying bricks to a kiln. The girls, knowing the old woman's reputation as a witch, stepped aside as much as they could to avoid any physical contact. Katharina's version of what followed was that she gave them a dirty look and wide berth but, because the path was so narrow, brushed their clothing and walked on. The girls' version was that one of them (whose mother owed money to Frau Reinbold) had been hit on the arm and that the pain in that arm had increased until she could no longer feel or move her hand. Katharina's enemies, including the girl's family, contrived to have Katharina brought before the bailiff, still Lutherus Einhorn. He called in a medical consultant, none other than Frau Reinbold's brother, who had earlier held his sword to Katharina's throat. Einhorn's verdict was, "It is a witch's grip; it has even got the right impression."

At this point, Katharina Kepler made a disastrous move. She attempted to bribe Einhorn with a silver goblet if he would proceed with her libel action and forget the arm incident. This was a windfall for Einhorn, still fearful that his part in the lawsuit would come to light. He suspended the libel case and sent charges of "witch's drink" and "witch's grip," as well as attempted bribery, to the High Council in Stuttgart. Christoph Kepler, Margarethe, and Margarethe's husband made a quick decision. Though there was a risk of implying that she had fled because of a bad conscience, they bundled Katharina off to Margarethe's house in Heumaden and from there to Kepler in Linz, just in time, for the council issued an immediate order for Katharina's arrest and "strenuous examination" about these matters and her theological beliefs. A witch trial had begun.

Katharina lived with Kepler and his wife in Linz for almost a year, until the following September, 1617. She was not a congenial presence. Kepler's description of her much earlier had not been flattering, but now that she was such an elderly woman—she was sixty-eight— he was ready to attribute her "trifling, nosiness, fury, and obstinate complaining" to old age. The household that year included himself and his wife Susanna, his two surviving children by Barbara—

Susanna and Ludwig—and one-and-a-half-year-old Margarethe Regina. In late spring there was a new baby, christened Katharina after her grandmother.

During that year, Kepler put his mind and efforts to refuting the charges against his mother and preserving his own safety and reputation. He hired lawyers for Katharina in Leonberg and for himself in Tübingen and Stuttgart, for there were rumors that he himself dabbled in the "forbidden arts." He wrote to the vice chancellor of the duke of Württemberg, informing him of Einhorn's bias in the case.

In September 1617, double tragedy struck the Kepler family. Two-year-old Margarethe Regina died, and the same month the news arrived of the death of her twenty-seven-year-old namesake, Kepler's stepdaughter Regina, the dearly loved child who had accompanied him and Barbara into exile and whom he had watched grow to womanhood during the happier years in Prague. Regina's husband Philip Ehem pleaded with Kepler to send his eldest daughter, fifteen-year-old Susanna, to Regensburg to care for the three motherless grandchildren.

Kepler and Katharina traveled with Susanna up the Danube from Linz to Regensburg, and then, after seeing Susanna settled there, journeyed to Württemberg. The interest in Katharina seemed to have abated. Kepler hoped he might get her libel suit back in motion, but that was a fruitless effort, as was a visit to Tübingen to try to reconcile himself with those who still thought him a closet Calvinist. He visited the very elderly Mästlin, and they discussed the forthcoming Rudolfine Tables at length.

It seemed safe enough to leave Katharina in Leonberg, so Kepler returned to Linz. He arrived home just before Christmas to discover that the six-month-old baby Katharina was ill. She died on February 9, 1618. Kepler's new wife was suddenly childless, and Kepler had lost three daughters within six months.

22

An Unlikely Harmony

1618–1627

KEPLER WAS TOO DISTRACTED with grief to concentrate on the tedious calculations required for the Rudolfine Tables. "Since the *Tables* require peace," he wrote, "I have abandoned them and turned my mind to developing the *Harmony*." The *Harmony* was a continuation of the book he had begun in Graz during the time when he and Barbara had mourned the death of his first infant Susanna. Now, in another profoundly heavy period, when the decimation of his family gave scant evidence of a rational, loving deity, he nevertheless returned to this attempt to reveal what he believed was the wondrous wisdom and rationality of God in nature. His research followed up on his conviction that mathematical harmonies among the planetary orbits, speeds, and distances from the Sun must be linked on a deep level with music. In 1607 Kepler had acquired a Greek manuscript by Ptolemy, also entitled *Harmony*, that had pre-empted his own ideas by about fifteen hundred years. He was both stunned and inspired by the similarity.

During Kepler's lifetime, evolving musical theory had added to the list of musical intervals that the ancient Greeks had declared pleasant to the human ear. There were now seven ratios that were accepted as the basis for what was called the "just" scale. Kepler had listened to these intervals and found he agreed with the additions. With his usual brand of curiosity, he wondered why God had chosen these numbers to produce musical consonance. Why leave out the number 7, for example? Some divisions of a harp string produce harmony, while an infinite number of others do not, and that reminded him of his insight that an infinite number of polygons could produce only five polyhedrons. He began to look for a similar way that the ratios of musical consonance had been singled out. He thought the "knowability" of the polygons might provide the answer.

Kepler began by dividing the polygons into levels of "knowability." The triangle, square, pentagon, hexagon, and octagon could all be constructed with ruler and compass, the classical Euclidean tools. Kepler dubbed them "knowable." Since the heptagon (seven-sided) couldn't be constructed with ruler and compass, he dubbed it "unknowable." Likewise nine- and eleven-sided polygons.

Kepler uncovered a mysterious link. If the number of sides of the *knowable* polygons (3, 4, 5, 6, and 8) were used in the ratios between string lengths, harmony resulted. For instance, both the triangle and the square were, by Kepler's definition, "knowable," and a ratio of string lengths of 3:4 produced a harmonious musical interval. The triangle and the pentagon were "knowable," and a ratio of string lengths of 3:5 produced a harmonious interval. On the other hand, a heptagon, with seven sides, was "unknowable." Sure enough, a ratio with a 7 in it produced dissonance. It seemed logical to Kepler that the numbers of sides in the unknowable polygons would have been avoided by God when designing the universe. Hence 7, 9, 11, and so forth were not part of ratios producing musical consonance. Kepler reasoned that because human beings are fashioned in the image of their Creator, they have an innate ability to enjoy manifestations of consonant ratios, an ability that doesn't require any knowledge or awareness of the mathematics or geometry involved. Tycho had thought similarly when he designed Uraniborg. A house built on the principles of harmony would be conducive to lofty thoughts and worthwhile study, even for those unaware they were living in such a structure.

When Kepler had devised his polyhedral theory and compared the results with the available data, he had been content with a margin of discrepancy that his faith in Tycho's observations had not allowed him to tolerate later when he wrote *Astronomia Nova*. He decided to revisit the polyhedral theory and investigate what other principles, in addition to the polyhedrons, God might have used in setting up the solar system, principles that could explain the discrepancies Kepler knew he now had to take more seriously. Kepler's research included acquiring an extensive knowledge of music theory, for he was becoming more and more convinced that the answers he sought were intimately connected with the combinations of musical intervals that human ears find pleasing.

Kepler examined the planets' distances from the Sun at perihelion and at aphelion, and their mean distances from the Sun. He could find no helpful harmonious relationships there. He tried looking for relationships between a single planet's slowest speed (at aphelion) and its fastest speed (at perihelion), and between and among those speeds using more than one planet. Within a few months he did indeed find an arrangement that was true both to the principles of musical harmony and to the planets' observed distances, speeds, and eccentricities.

Of more significance, on May 15, 1618, as he was finishing the book, he discovered a third law of planetary motion, his "harmonic law," the true relationship between the orbital periods of the planets and their distances from the Sun. Kepler was ecstatic, wanting to give way to a "sacred frenzy," as he put it. "I am . . . writing the book," he rejoiced, "whether for my contemporaries or for posterity, it does not matter. It can await its reader for a hundred years, if God Himself waited six thousand years for His contemplator." Near the end of the book he included a prayer that vividly reveals this remarkable man:

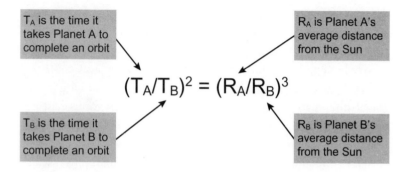

T_A is the time it takes Planet A to complete an orbit

R_A is Planet A's average distance from the Sun

$$(T_A/T_B)^2 = (R_A/R_B)^3$$

T_B is the time it takes Planet B to complete an orbit

R_B is Planet B's average distance from the Sun

Figure 22.1: Kepler's third law of planetary motion, the "harmonic law." Kepler discovered the true relationship between the orbital periods of the planets and their distances from the Sun in 1618, as he was finishing his book *Harmonice Mundi*. Kepler's third law of planetary motion states that the ratio of the squares of the orbital periods of two planets is equal to the ratio of the cubes of their average distances from the Sun.

O you who by the light of nature arouse in us a longing for the light of grace, so that by means of that You can transport us into the light of glory: I give thanks to You, Lord Creator, because You have lured me into the enjoyment of Your work, and I have exulted in the works of Your hands: behold, now I have consummated the work to which I pledged myself, using all the abilities that You gave to me; I have shown the glory of Your works to men, and those demonstrations to readers, so far as the meanness of my mind can capture the infinity of it, for my mind was made for the most perfect philosophizing; if anything unworthy of Your deliberations has been proposed by me, a worm, born and raised in a hog wallow of sin, which You want mankind to know about, inspire me as well to change it; if I have been drawn by the admirable beauty of Your works into indiscretion, or if I have pursued my own glory among men while engaged in a work intended for Your glory, be merciful, be compassionate, and forgive.

Kepler dedicated the five-volume work *Harmonice Mundi* (Harmony of the World) to King James I of England, expressing the hope that these examples of the glorious harmony with which God had endowed his creation might strengthen James in attempts to bring harmony and peace among the tragically divided churches and other polities. However, only four days before Kepler completed his book and penned that dedication, Protestant Bohemia, where he had spent the ten best years of his life, exploded in a revolution that began the Thirty Years War.

For Kepler there was soon to be trouble closer to home. In the summer of 1618, an ominous letter came from an old classmate (possibly acting as Kepler's lawyer) on the law faculty at the University of Tübingen, warning of a strategy that the Reinbolds and Einhorn might be planning. In autumn 1619 the warning proved correct. A counter civil suit was filed against Katharina Kepler for damages for poisoning Frau Reinbold with the "witch's drink." By that time Katharina's enemies had collected a forty-nine-count indictment against her, including a plethora of local gossip and fancy that recalled unnatural, eerie behavior. The charges included riding a calf to death, muttering fatal "blessings" over infant children, causing pain without touching people, the unnatural death of animals, and trying to entice a young girl to become a witch. One accusation was true. Katharina, having heard in a sermon about an archaic custom of fashioning goblets from the skulls of dead relatives, had asked the gravedigger for her father's skull so that she could have it set in silver for her son Johannes, the imperial mathematician.

With Einhorn still acting as bailiff, testimony began in November 1619. The following July the Reinbolds succeeded in getting the duke of Württemberg to turn their complaint into a criminal case. A few days later, on August 7, the seventy-four-year-old Katharina was awakened from her sleep in the dead of night, bundled into a large chest, carried out of her daughter's house, and put in prison in chains. At this point her son Christoph managed to have the trial, with all the spectacle and scandal attached to it, transferred to Güglingen, but he and Margarethe's husband, Georg Binder, were inclined to abandon Katharina and scramble to salvage whatever they could of their own dwindling reputations. The faithful Margarethe was of a different mind. Once again she wrote to her brother in Linz. Kepler applied to the duke of Württemberg for a delay in the trial until he could arrive, for he planned to defend Katharina himself.

Kepler chose to take his family with him when he left Linz that September of 1620. He and Susanna now had a young son, Sebald, who had been born in January 1619, and Susanna was pregnant again. They crept away like thieves in the night, without even telling Kepler's assistant Gringalletus where they were headed. The reason for the trip was too shocking to have it spread abroad in the town. Kepler left his family in Regensburg, where his daughter Susanna may still have been living with Regina's family, and proceeded alone on his grim journey to Württemberg. The mystified people of Linz thought their mathematician had fled for good.

Kepler found his mother in prison in chains with two guards and required to pay these guards herself, as well as for her food and upkeep. The Reinbolds complained that so much of her money was being used up in this manner that there would be little left for them when the trial was over.

Kepler had been advised that having the defense case written down would help the outcome, and he insisted that all the defense lawyer's arguments be put in writing. Christoph lamented the greater cost for what he thought was already a lost cause. The proceedings dragged on, with more lawyers, more witnesses, and more written arguments. Kepler traveled to Stuttgart to consult his lawyer in person, and they put together a 126-page legal brief, much of it in Kepler's handwriting, that rebutted the charges one by one. The trial ended in August, and all the proceedings were sent, as was the custom, to the law faculty of the University of Tübingen. It was they who would make the decision. Kepler's friend Christoph Besold was on that faculty. Nevertheless, even the force of Kepler's presence throughout the trial, his skill in devising the defense, and his powerful friend could not bring about an acquittal. The court declared itself uncertain and ordered that Frau Kepler be examined once more under the lightest form of torture, verbal terror while being shown the instruments of torture.

On September 28, 1621, Katharina was dragged to the torture chamber, accompanied by three representatives of the court, a scribe, and a bailiff (not Einhorn this time). The torturer himself showed her his instruments, described their use, and with the greatest possible sternness and melodrama commanded her to tell the truth. Contrary to all expectation, Katharina Kepler gathered her aging wits about her, summoned the eloquence she had bequeathed to her son, and saved herself. As the report reads:

> She announced one should do with her what one would. Should one pull one vein after another out of her body, she knew that she had nothing to say. She fell to her knees, uttered the Lord's Prayer, and declared that God should make a sign if she were a witch or a demon or ever had anything to do with sorcery. Should she be killed, God would see that the truth came to light and reveal after her death that injustice and violence had been done to her, for she knew that He would not take His Holy Spirit from her but would stand by her.

The charges were dismissed, and Katharina was set free. The Reinbolds were fined ten florins for having begun the proceedings, and Christoph Kepler was ordered to pay thirty florins for the expense of transferring the trial to Güglingen. Despite the power of her self-defense, Kepler's mother was a broken woman, and she died the following April.

As soon as he knew his mother was acquitted, Kepler set off on the journey back to Linz. There were two young children in his family now, for Susanna had given birth to a daughter, Cordula, the previous January, and Sebald was nearly three. But Linz, in the autumn of 1621, was a different city from the one they had left in secret a year earlier. Not long after his departure, the Bohemian rebel army had been defeated at the Battle of White Mountain. The Protestant revolution in Bohemia was over, but the Bavarian army still occupied Linz. In Prague, Ferdinand II, the same man who had been responsible for the Counter-Reformation edicts that had forced Kepler out of Graz years before, was now emperor of the Holy Roman Empire and was overseeing the brutal execution of Protestant leaders. Jesensky (who had aided Kepler in his early unhappy negotiations with Tycho, had delivered the eulogy at Tycho's funeral, and had been Kepler's powerful friend during the years in Prague) was one of them. Jesensky's tongue was cut out before he was quartered. His gory head, along with others, was stuck on a pike on the bridge tower, where they were left to decompose for ten years and finally fell off, one by one, onto the bridge or into the river.

There was fortunately less savage treatment of Protestants in Linz. Kepler remained unscathed, despite speculation in the city that he had fled because Ferdinand had put a price on his head, an odd suspicion since everyone knew the Lutherans had excluded Kepler from Communion. Though Kepler entertained some doubts that it would happen, Ferdinand reconfirmed his appointment as imperial mathematician. The next year, 1622, when all Protestants in Linz were required to convert to Catholicism or leave the city, Kepler was allowed to stay, although his library was sealed for a time and he agonized over the requirement that to have it unsealed he must choose which of his beloved books to surrender to the censors. His children were forced to attend Catholic church services. But he was allowed to keep Protestant Planck, his printer, with him along with as many skilled assistants as Planck required. All the turmoil had not completely halted Kepler's

A seventeenth-century drawing showing the heads of executed Protestant leaders impaled on the bridge in Prague.

work on the Rudolfine Tables, nor had the death of yet another of his children, four-year-old Sebald, in the early summer of 1623.

The production of the Tables had been a matter of the highest priority when Tycho died, and during his years in Prague Kepler had filled hundreds of sheets with calculations in preparation for eventually completing them. Kepler's discovery of his planetary laws were a huge step toward this end, but those discoveries also presented new challenges.

Kepler had derived his first two laws from the Mars observations. In order to complete the Tables, he had to show that the same laws applied to the other planets. Much of this work had been accomplished in connection with the writing of *Epitome* and *Harmonice Mundi,* but it was not finished. Kepler explained in the preface to the Tables that the reason for the long delay in their appearance was "the novelty of my discoveries and the unexpected transfer of the whole of astronomy from fictitious circles to natural causes." No one, he pointed out, had ever attempted anything of the kind before.

In 1617, while at work on *Harmonice Mundi*, Kepler had come across a book by John Napier on logarithms. A year later he realized how much this new invention would simplify the computations that took so much of his time. In the winter of 1621–22 he wrote his own book on logarithms, and he proceeded to use them to solve some of the problems involved in composing the Tables.

At last, in 1624, Kepler finished the Rudolfine Tables in the new logarithmic form. Tables like these did not give daily positions of the planets; rather, they were far more generally useful, making it possible to figure out any planet's position for any time thousands of years into the future or the past. In the case of the Rudolfine Tables, Kepler's instructions about their use, with examples, took up about half the volume of the work. Kepler included logarithm tables, Tycho's catalog of a thousand stars, and latitudes and longitudes of many cities.

Once again, having completed a major work, Kepler's difficulties with it were far from over. Financing publication was a serious problem. A trip to Vienna, where the imperial court was now situated, won him promises that payment would be forthcoming via various cities that would turn imperial funds over to Kepler, but after ten months of traveling among these cities, he came back with virtually nothing to show for the effort. He had taken the opportunity to order four bales of paper from the cities of Memmingen and Kempten and had them sent directly to Ulm, where he expected to print the book. Eventually he would have to pay for the publication himself, with Tycho's heirs all the while trying to claim a share of the profits and censorship rights. Nevertheless, it was the Rudolfine Tables, more than any of Kepler's other works, that led to the widest recognition of what he had achieved.

A manuscript such as this, with 120 pages of text and 119 pages of complicated tables, was a printing challenge beyond anything Planck had attempted so far, and he, in any case, was eager to leave Linz and its religious turmoil. Ulm seemed to Kepler the best place to print the book, where there were other skilled printers and no war going on, but Emperor Ferdinand rebelled at the notion of having the work done outside Austria.

As this was debated, the Thirty Years War, which had not ended with the quashing of the Bohemian rebellion, came perilously close again. A peasant uprising in the summer of 1626 almost succeeded in driving the Bavarian troops and Ferdinand's forces out of Linz and Upper Austria. Kepler's house, situated on the city wall, had to be opened for soldiers guarding the wall, while peasant bands, burning and looting, threatened the capital. During this two-month siege, Kepler almost lost the Rudolfine Tables. On June 30, a fire started by peasant rebels spread and consumed Planck's press but somehow spared the handwritten manuscript.

The loss of the press and the near destruction of his manuscript were the last straw for Kepler. He had lived in Linz for fourteen years, longer than he had in Prague. He was exhausted by the confusion and disorder and wanted nothing so much as to complete the publication of the Rudolfine Tables in relative peace. When the siege was finally lifted in August, he wrote to request again the emperor's permission to depart, and this time it was granted. In mid-November 1626, the Kepler family took a boat up the Danube in the direction of Ulm. By this time, there were two more young children, Fridmar, three years old, and Hildebert, one. Their elder sister Cordula was five. Beyond Regensburg the river was completely frozen, so Kepler left his wife and children there and went on alone overland toward Ulm "on a wagon laden with plates of my figures and Table work." When he arrived, on December 10, 1626, he found lodgings across the street from the printing shop of Jonas Saur, who finally printed the Tables.

Kepler oversaw every aspect of the printing and worked almost daily with the typesetters. He had brought with him his own set of numerical type (which had not been destroyed when the press burned) and the astronomical type that he had had custom-made for the *Tables*. As the pages came off the press, he proofread each one. Kepler saw this book as the crowning achievement of two lifetimes, Tycho Brahe's and his own. Even though he had written it himself, discovered the new laws that made it correct, and done all the calculations, he put Tycho's name first on the title page as the primary author.

Kepler decided that the *Rudolfine Tables* should have an elegant frontispiece. He had an idea in mind and asked a friend from Tübingen, Wilhelm Schickard, to prepare a sketch of it. The frontispiece summed up Kepler's concept of the world of astronomy, including its history, and was at the same time a masterpiece of whimsy. It shows a pavilion with twelve columns. Those at the back are hewn logs, and a Babylonian astronomer stands there using only his fingers to make an observation. Babylon was where astronomy had its roots. Nearer the front, Hipparchus on the left and Ptolemy on the right stand by columns built of brick. Closer in the foreground sits Copernicus by an Ionic column on whose pedestal he has propped his famous book, and Tycho stands by a Corinthian column with some of his celebrated instruments hung on it. He and Copernicus are deep in discussion, presumably about the Tychonic and Copernican systems, for Tycho points at the ceiling of the temple, where there is a drawing of his system. Kepler cunningly has him not telling Copernicus that it is correct, but asking, "Quid si sic?" (How about that?)

Ringing the rooftop are six goddesses, each a symbol of something that helped Kepler in his discoveries: Magnetica (on the far right); then Stathmica, the goddess of law; Geometria; Logarithmica; and finally a goddess holding a telescope and another with a globe that casts a shadow.

The frontispiece of the *Rudolfine Tables.*

At the very top of the pavilion flies the Hapsburg eagle, with coins dropping from its beak, a symbol that needs no explanation.

Kepler did not show Tycho's heirs the panels in the base of the pavilion before publication, though they would have approved the center panel, a map of Hven. To the left is a panel showing Kepler sitting at a table, by candlelight, a few numbers scratched on the table-cloth, his major books listed on a banner above his head, and a model

Detail from the frontispiece of the *Rudolfine Tables*.

of the roof of the temple on the table before him. Tycho stands above beside the most elaborate column, but it is Kepler who has labored in the basement, at night, and brought about this marvelous achievement, this temple of the goddess of astronomy Urania, the *Rudolfine Tables*. Very few of the coins are dropping onto Kepler's desk.

The *Rudolfine Tables* lived up admirably to Tycho's and Kepler's hopes for them. The planetary positions given by the *Tables* were much more accurate than those given by the Alfonsine or Prutenic Tables or tables that had been composed by Longomontanus and others. Predictions for Mars, for instance, had previously erred up to five degrees. The *Rudolfine Tables* stayed within plus or minus ten *arcminutes* of the actual positions.* In 1629, when Kepler was preparing an ephemeris for the year 1631, he realized that because of the depend-

*This information comes from Owen Gingerich (1973).

ability of his *Rudolfine Tables,* he could confidently predict two "transits" that would occur during that year—one of Mercury and another of Venus—across the disk of the Sun.* He published his predictions in a short pamphlet, *De Raris Mirisque Anni 1631 Phenomenis* (1629). He would not live to see how superbly accurate he had been.

---

*On November 7, 1631, the astronomer Pierre Gassendi observed the Mercury transit from Paris. The result was a triumph for Kepler's astronomy. The transit of Venus was not visible in Europe, because it was night there when it occurred.

23

MEASURING THE SHADOWS

1627–1630

ONCE BEGUN, printing the *Tables* went quickly. In early September 1627 Kepler took copies to the Frankfurt Book Fair and finally rejoined his family in Regensburg in early December, only to leave them again after Christmas to take a presentation copy of the book to Emperor Ferdinand. The court was in Prague, where Ferdinand was installing his son as the king of Bohemia, and everyone was in exceptionally good spirits because the Protestant revolt had finally been completely put down. The most immediate cause for celebration was the defeat of invasions in the north by the Protestant King Christian IV of Denmark, none other than Tycho Brahe's old nemesis. Christian had been driven from German soil and also from the entire formerly Danish peninsula of Jutland.

Noticeably absent were all Kepler's old Protestant friends, including poor Jesensky, whose head was a grisly presence on the bridge tower. But many other old friends were in Prague as well as many admirers, and the emperor was so pleased with the *Rudolfine Tables* that he granted Kepler four thousand florins, ten times his yearly salary. That brought to twelve thousand florins the amount of money owed him from the treasury, or the equivalent of thirty years' salary. Kepler knew he would never collect that if he left the emperor's service. As it was, he was told that all his fears that he had already lost his job because of the emperor's edicts in Linz were groundless. All he needed to do was convert to Catholicism. Kepler, of course, refused.

Kepler had been offered a job in England, and he might at this point have made an abrupt decision to forfeit all back salary and go there immediately had it not been for a man with whom Kepler had had a long association by correspondence but not met in person before. Albrecht Wallenstein had commissioned a horoscope from Kepler in 1608 without revealing who he was or anything about himself except the date and time of his birth. Kepler, always good at horoscopes, had produced one that greatly impressed Wallenstein.

Wallenstein was a favorite of the emperor and was, in fact, the general largely responsible for the defeat of Christian IV of Denmark. He let it be known that he believed the different faiths must coexist peacefully, and he allowed the practice of Protestantism in his Silesian duchy of Sagan. By moving there, Kepler would be able to maintain his faith while remaining in imperial service. As the agreement finally was worked out the following February 1628, Kepler was promised a house, a printing press, and a generous stipend of a thousand florins a year.

Kepler made a final trip to Linz in the early summer, where the beleaguered city surprised him with a payment of two hundred florins for their presentation copy of the *Rudolfine Tables*. Kepler moved his family to Sagan in July. He was fifty-six years old.

Kepler was unhappy in Sagan. No one knew him, and he knew no one. There was little intellectual stimulation, and the local dialect was so different from the German Kepler spoke that he had difficulty understanding it or making himself understood. He was suffering, as usual in his later years, from eczema and abscesses. Most discouraging of all, he had hardly arrived when the Counter-Reformation followed him. Once again he was exempted from enforced conversion

Kepler's daughter, Susanna Kepler Bartsch, in a painting done in 1630.

and the banishment of those who refused to convert, but it was a bitter experience to see it all happening for a third time.

In the winter and spring of 1630, Kepler's mood lifted a little. The
long-promised printing press and a printer to work at it finally materialized, saving Kepler the trouble of setting his own type by hand
and taking it to a nearby town for printing. In March Susanna,
Kepler's daughter by Barbara, married Jacob Bartsch, a student of
mathematics and medicine in Strasbourg who had worked for Kepler
as an assistant. Kepler decided that the wedding should take place in
Strasbourg, though that was too far away for him to attend, given his
age and the fact that his own wife would be eight months pregnant
by the time of the festivities.

Matthias Bernegger, a longtime friend and correspondent living
in Strasbourg, had recommended Bartsch as a suitor and brought the
young couple together. Bernegger gave Kepler a detailed and glowing
account of the celebration in a letter. On March 12, the bridegroom
received his medical degree in the morning, and later in the day the
couple were wed. Kepler's brother Christoph, his sister Margarethe,
and his son Ludwig were part of a wedding procession that included

Strasbourg's most prominent citizens, and huge crowds lined the streets as they passed. "It was meant, especially, to honor you," Bernegger told Kepler.

A month later Kepler's youngest child, Anna Maria, was born. Now he had two grown children by Barbara and four younger ones by Susanna, although two of those would not live to adulthood. There had been five others who had died in infancy or early childhood. There had been Friedrich, the shining boy who had died in Prague, and Kepler's beloved stepdaughter Regina, who had died as a young married woman.

In the days just after Anna Maria's birth, when Kepler could not leave his wife's side long enough to oversee the printing of his next series of ephemerides, he told his printer to work instead on a book that he had begun as an essay when he was a student in Tübingen. During the Prague years he had expanded the essay into a short story, much to the delight of his friends, for he laced it with puns and allusions that they could appreciate. This was, in fact, the piece that he had feared might jeopardize his mother's case in the witch trial, and with the trial long past he felt vindicated in publishing it, with some notes pointing out how it might have been used out of context. *Somnium* (The Dream) consists of a twenty-eight-page story and fifty pages of notes and diagrams and is widely regarded as the first work of science fiction. The hero and narrator, in the course of his travels, visits Uraniborg, where he finds Tycho and many assistants all speaking languages he does not understand until he learns sufficient Danish. Most of the story takes place on the Moon and is about the way the heavens and Earth appear to inhabitants there.

The notes concerned much more than the possible misuse of the story in the witch trial. It was here that Kepler showed that he understood the concept of gravity better than he is often given credit for. He described clearly the point between Earth and Moon where their separate gravitational attractions exactly cancel out.

The printing of *Somnium* proceeded sporadically between print-

ing runs for the ephemerides. It was not finished when Kepler set out
from Sagan in October to try, once again, to recover some of his back
salary. This time it seemed there was hope. After being put off again
and again, Kepler had been promised that if he appeared in Linz on
November 11, he would be paid some interest on investments there.
He also planned to attend a meeting taking place near that time in
Regensburg at which the future of his patron Wallenstein, now fallen
from grace, hung in the balance.

Kepler was exhausted when he rode out of Sagan on October 8.
He had pushed the printer and himself unmercifully. He had needed
to ship ahead of him, for the autumn book fair in Leipzig, a large
stock of books: fifty-seven copies of ephemerides, sixteen copies of
the *Tables,* and seventy-three other books. It had also taken time and
effort to gather together all the documentation he had collected
through the years about everything that was owed him.

After Leipzig, Kepler rode to Regensburg and arrived there on
November 2 after a cold autumn journey on a nearly worthless old
horse that he sold in the city for a few florins. He felt ill but, determined not to neglect anything he had set out to do, shrugged off his
illness as no more than a nuisance. For a man his age, it was more
than that. He grew worse, his fever soared, and he lapsed into delirium. A doctor came and bled him, which did not help.

At last a Protestant pastor was summoned. Kepler drifted in and
out of consciousness for several days and in his few lucid moments
tried to explain to the pastor that he had done his utmost to recon-
cile Catholics and Protestants. The pastor admonished the dying
Kepler that this was like expecting to bring together Satan and
Christ. Kepler, as usual, was not dissuaded from his own beliefs.
When asked on what basis he hoped for salvation, he answered,
"Solely on the merit of our Savior Jesus Christ, in whom is found all
refuge, solace, and salvation." Those words were the very essence of
Protestantism, and no pastor could take exception to them.

On November 15, 1630, Kepler died. Though his grave has been lost, he was not buried in obscurity. Some of the most powerful and illustrious men of the empire were in Regensburg for the meeting Kepler had planned to attend, and many of them walked in the funeral procession for the imperial mathematician they had so celebrated but also so poorly supported. That evening there was a meteor shower. As it was reported at the time, fiery balls fell from heaven.

Kepler's gravestone in the Protestant cemetery bore an epitaph that he had written himself:

> I measured the heavens, Now the earth's shadows I measure,
> Skybound, my mind. Earthbound, my body rests.

Appendix 1

ANGULAR DISTANCE

A simple way to approach the definitions of the terms *angular distance, angle of separation,* and *degree of arc* is to imagine oneself at the center of a giant clock face, where the two hands meet. From that point of view, the angle of separation or angular distance between an object at "twelve o'clock" and an object at "one o'clock" is thirty degrees of arc. Likewise the angular distance between "one o'clock" and "two o'clock," and so forth. The angular distance between an object at "twelve o'clock" and another at "two o'clock" is sixty degrees of arc, and so forth. The entire circle has 360 degrees. To understand the concept roughly with regard to the sky, draw an imaginary line between two stars directly overhead whose angle of separation you want to measure and let that line continue all the way around you and the earth beneath your feet until it comes up the other side of the sky and joins its other end, so that the line has drawn a huge circle all the way around the celestial sphere. That huge circle is the equivalent of the clock face, and you are in the center where the two hands meet. If the two stars look to be at, let us say, one and two o'clock, then their angular separation is thirty degrees.

Because celestial objects that interest astronomers are often closer together than one degree of arc, degrees are divided into smaller segments. There are sixty minutes of arc in one degree of arc; sixty seconds of arc or arcseconds in one minute of arc.

Two objects whose angle of separation is, let us say, thirty degrees of arc (from twelve to one on the clock face as viewed from the center of the clock) can actually be either quite close to one another or very far apart. For example, looking from the window of my study, I see two trees whose distance from one another (if you go out and measure it) is about twelve feet. Their angular separation from where I stand is about thirty degrees. Beyond them is the sky. Lining up each tree with a star, those two stars also have an angular separation from one another of about thirty degrees when viewed from my study. However, those two stars are definitely not just twelve feet apart. Knowing what angle separates two objects does not tell us the *distance* between them.

Appendix 2

VOCABULARY OF ASTRONOMY

Much of the vocabulary that is essential to understanding this book is explained in the relevant chapters, but here are a few more useful terms:

Meridian circle: Starting at the north celestial pole, draw an imaginary line to the zenith above where you are standing, then continue the line around the celestial sphere until you have brought it all the way around the celestial sphere, through the south celestial pole, to meet its tail again at the north celestial pole. What you have drawn is a line of longitude or a meridian, the celestial equivalent of the lines of longitude or meridians to be found on a globe of Earth. This meridian is perpendicular to the horizon.

Altitude is the distance of a star or planet above the horizon, measured in degrees. A complete circle is 360 degrees, so the altitude of a star at the zenith is 90 degrees. No star can ever have an altitude greater than 90 degrees.

Azimuth is the distance of an object from the meridian, also measured in degrees. Imagine again drawing the meridian hoop. Stand facing north and imagine that line. If you see a star off to the left or right of that line, that star is not on your meridian. Its azimuth is the measurement of how far it is from the meridian.

Meridian, altitude, and azimuth—like horizon and zenith—are dependent on where an observer is standing.

Astronomers need measurements that will not change with the position of the observer—measurements that stay put, as do the celestial equator and the celestial poles. An astronomer in Denmark must be able to tell an astronomer in Italy what the position of a star or planet is without using Denmark as a reference point. Hence another set of terms:

The *prescribed meridian* is the meridian line established not by the position of the observer but by the position of the Sun at the vernal equinox.

The *declination* of a celestial object is its distance in degrees above the celestial equator.

Right ascension is its distance in degrees east of the prescribed meridian.

Declination and right ascension are hence independent of the observer's position on Earth. Whether you are in New York or Arizona or Turkey, the declination and right ascension of a particular star will be the same.

Two more measurements are related not to the horizon or the celestial equator but to the ecliptic:

The *latitude* of a celestial body is how many degrees it is above or below the ecliptic.

Longitude is a body's position *along* the ecliptic, measured in degrees eastward from the vernal equinox.

It is useful to remember all these terms in groups of four:

Altitude and *azimuth* are measurements related to the *horizon* and the *meridian*.

Declination and *right ascension* are measurements related to the *celestial equator* and the *prescribed meridian*.

Latitude and *longitude* are measurements related to the *ecliptic* and the *vernal equinox*.

APPENDIX 3

KEPLER'S USE OF TYCHO'S OBSERVATIONS OF MARS
TO FIND THE ORBIT OF EARTH

To discover what Earth's motion was like, Kepler put himself and the readers of *Astronomia Nova* in the position of a Martian astronomer observing Earth.

The Martian begins the series of observations when Mars is on Earth's apsidal line (that is, the line running through the Sun, the center of Earth's orbit, Earth, and Earth's positions at aphelion and perihelion—see figure 19.1). Every 687 Earth-days after that (687 Earth-days is one Martian year), the Martian takes another observation of Earth. Each time, Mars has completed an orbit and returned to Earth's apsidal line. To put himself and his readers in the place of that Martian observer, Kepler reversed the direction along which Tycho had observed Mars from Earth, in effect allowing himself to watch Earth from a stationary Mars.

Appendix 3, Figure 1 shows the Sun and Earth and Mars arranged so that Mars is at opposition, but the diagram gives no indication of distances. It shows only that at this moment an observer on Earth (were it possible to see the stars in the sky at the same time as the Sun) would find the Sun at the point called Z-1 in the zodiac belt and Mars at Z-2; an observer on the Sun would find Earth and Mars at Z-2; an observer on Mars would find Earth and the Sun both at Z-1. At the moment captured here, Earth, Sun, Mars, and those points in the zodiac are all located on the same straight line; if Earth is on its apsidal line, then obviously so is Mars.

Using Tycho's collection of observations as his background data, Kepler found instances 687 Earth-days (a Martian year) apart, beginning when Mars was on Earth's

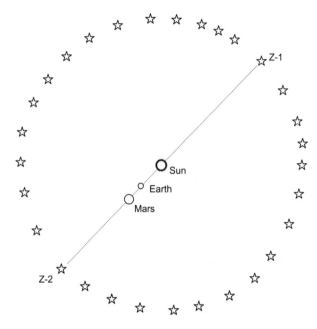

Appendix 3, Figure 1: The drawing represents a large circular room, with the stars of the zodiac "belt" painted on its walls. These stars, whose positions and angular distances from one another were well known to early astronomers and astrologers, are the fixed background against which an inhabitant of the Solar System observes the planets. According to Copernicus, the Sun is in the center, and Earth is near it. The stars are, of course, much farther away from the Sun and Earth than the dimensions of a room can possibly simulate. But their great distance means that whether viewed from Earth, the Sun, or Mars, they occupy the same positions in the zodiac, just as though they really were on the walls of a huge room.

apsidal line. Though Mars would have arrived back on that line at the end of every 687-day period, Earth—which takes only 365 days to complete an orbit—would not have. Earth would be in a different place each time. Appendix 3, Figure 2 imagines Earth showing up at positions designated Earth^1, Earth^2, and Earth^3, while Mars's position is always on Earth's apsidal line. Mars, Sun, and Earth^1 are points of a triangle; Mars, Sun, and Earth^2 are points of another triangle; Mars, Sun, and Earth^3 are points of a third triangle. All the triangles have a side in common, the Mars-to-Sun line. The length of that line (which coincides with Earth's apsidal line) is the same in all three triangles.

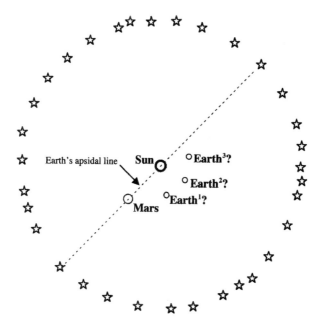

Appendix 3, Figure 2: At the end of each 687-day interval, Mars has completed an orbit and made its way back to Earth's apsidal line, while Earth (requiring only 365 days to complete an orbit) shows up at different positions. In each case, Mars, the Sun, and Earth are points of a triangle. The three triangles have one line in common, the line that coincides with Earth's apsidal line—the Sun-to-Mars line.

From this "Martian astronomer" exercise, Tycho's solar theory, and his own Vicarious Hypothesis, Kepler had the information needed to find out where observers on any of the three bodies (Earth, Mars, Sun) would see the other two bodies against the background stars of the zodiac, when Earth was at Earth[1], Earth[2], and Earth[3] (see Appendix 3, Figure 2).

Knowing the angular distances between the positions where the three lines illustrated in Appendix 3, Figure 3 ended in the zodiac gave Kepler the angles of the triangle when Earth was at Earth[1], which in turn told him how the lengths of the sides were related to one another. He made this calculation for the triangles when Earth was at Earth[1], Earth[2], and Earth[3]. All three triangles had one side in common—the Sun-Mars line (line C in Appendix 3, Figure 3), which coincided with Earth's apsidal line. Comparing the lengths of the other sides with the length of that common side

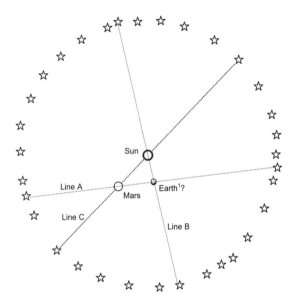

Appendix 3, Figure 3: The triangle when Earth was at Earth[1]: *Tycho's observation,* made from Earth, of Mars's position in the zodiac told Kepler where the Earth-Mars line (line A) ended in the zodiac. *Tycho's solar theory* (the theory in which the Sun orbits the Earth and the planets orbit the Sun) gave what Kepler judged to be accurate positions for the Earth in the zodiac as viewed from the Sun, which means he knew where the Earth-Sun line (line B) ended in the zodiac. *Kepler's Vicarious Hypothesis's heliocentric longitudes* told him where Mars appeared in the zodiac when viewed from the Sun; i.e., where the Mars-Sun line (line C) ended in the zodiac.

told him how *all* the sides of all three triangles compared with one another. And that knowledge, in principle, allowed Kepler to find the position of Earth in its orbit at the time each of the observations were made—Earth[1], Earth[2], and Earth[3]—in other words, to draw a diagram placing the three Earth positions where they really do occur, not just in imaginary places. Using those three points, he could draw a circle through them to represent Earth's orbit and find out where the center of the circle was. That should give him the position of the center of the orbit relative to the Sun and the radius of the orbit. Kepler double-checked his findings using several different sets of triangles based on different observations.

Drawing all those points in turn might seem certain to have described the shape of Earth's orbit and to have revealed that it was an ellipse. To think so is to underestimate the pitfalls and uncertainties involved, and the subtlety of the answer Kepler was seeking. The ellipse he would later discover is so close to being a circle that it was impossible to find it by this method, even had there been no error at all in his calculations. For this particular procedure, each piece of data provided an opportunity for error, and in a process that used three triangles with one common side, that meant seven chances to go wrong. The trigonometry used in the computation, especially when it involved small angles, magnified any error. Even if the true orbit *had* been circular, Kepler might easily have got different results for different sets of triangles. To his chagrin, he did.

However, this outcome was by no means a complete disappointment, because Kepler's results showed that contrary to what Ptolemy, Copernicus, and Tycho had thought, the center of Earth's orbit lay somewhere in the middle between the equant point and the Sun, and that was where astronomers had traditionally put the center of the orbit of a *planet*. Also, Kepler had found that Earth was *moving* like a planet, speeding up when it came closer to the Sun and slowing down as it moved away.

NOTES

JKGW refers to the twentieth-century compendium of Kepler's works and letters: Max Caspar, Walther von Dyck, Franz Hammer, and Volker Bialas, eds. *Johannes Kepler Gesammelte Werke.* 22 vols. Munich: Deutsche Forschungsgemeinschaft, and the Bavarian Academy of Sciences, 1937–.

TBDOO refers to Tycho Brahe's collected works: John Lewis E. Dreyer, ed. *Tychonis Brahe Dani Opera Omnia.* 15 vols. Copenhagen: Libraria Gyldendaliana, 1913–29.

Mechanica refers to Tycho Brahe, *Astronomiae Instauratae Mechanica,* Raeder et al. translation.

I have used the following form for quotations where the original is in Danish and I have taken the translation from an English source: Original source/English translation source. For example: Gassendi 3:20/Thoren 20 means that the original is Gassendi, volume 3, page 20; and I have used the translation on page 20 of Thoren.

PROLOGUE

3   "You will come": Brahe to Kepler, Jan. 26, 1600, JKGW, vol. 14, letter 154.

1. LEGACIES

For much of the information in the early chapters of this book I am indebted to Victor Thoren, whose *Lord of Uraniborg* (1990) superseded Dreyer's 1890 book

Tycho Brahe: A Picture of Scientific Life and Work in the Sixteenth Century as the definitive biography of Tycho Brahe. Tycho provided an autobiographical summary in his *Mechanica,* 106 ff; and Christianson has done a splendid job of recreating Tycho's youth in *Cloister and Observatory: Herrevad Abbey and Tycho Brahe's Uraniborg.*

8   "without the knowledge": The quotation and Tycho's telling of the incident are from *Mechanica,* 106.

11   "was sent to grammar school": Ibid.

13   where Tyge lodged: For Thoren's speculation on this subject see Thoren 1990, 9–10.

13   grounded in the "liberal arts": On Philippist university curriculum, see ibid., 11.

2. ARISTOCRAT BY BIRTH, ASTRONOMER BY NATURE

25   chose the University of Leipzig: Tycho reported the move to Leipzig in *Mechanica,* 106, but placed it two years earlier and said nothing about the previous three years in Copenhagen.

27   "bought astronomical books" and following quotations: *Mechanica,* 107. Pages 107, 108 are Tycho's account of his early attempts in astronomy.

28   "rectify this sorry state of affairs": Thoren 1990, 17.

28   cross staff, or radius: *Mechanica,* 108.

28   "stayed awake the whole night": Ibid.

28   "had no opportunity": Ibid.

31   each demanding that the other draw his sword: The story of the duel was passed down by word of mouth over the next hundred years and finally written down by one Jacob Stolterfoht, a Lutheran clergyman who was the grandson of the woman who knew Danish. See Thoren 1990, 23. Tycho's own mention of the duel story is in *TBDOO,* 1:135–36.

32 (fn) Tycho's first biographer: Thoren 1990, 25/Gassendi, 10:209.

34   a new pair of compasses: *Mechanica,* 80–83.

34–35 "placing the vertex": Ibid., 107.

35   The *quadrans maximus:* Ibid., 88–91.

37   Ramus, in his next book: Petrus Ramus, *Defensio pro Aristotele adversus Iac. Schecium* (Lausanne, 1571).

38   "He dwells on earth": *TBDOO,* 9:173/Thoren 1990, 45.

3. BEHAVIOR UNBECOMING A NOBLEMAN

39   complicated system of reciprocity: For the feudal system in Denmark, see Christianson 2000, 25–26. Christianson has brought

together from many sources, including Scandinavian ones (see his bibliography), information about life in sixteenth-century Denmark with which he has been able to flesh out the sometimes scarce biographical information about Tycho.

40   Herrevad Abbey: The description of Herrevad and its background come from my own visit there, and from a thorough background and description of what it was like in Tycho's time, including much information about Steen Bille, in Christianson 1964.

41–42   "Verily, there did once": See Christianson 1964, 43.

43   Kirsten Jørgensdatter: For discussion of and speculation about her see Thoren 1990, 45–48, and Christianson 2000, 10–14. When I visited Knutstorp in 2000, I found the conviction still strong in the community that Kirsten had been the pastor's daughter.

43   "a woman of the people": Gassendi/Thoren 1990, 45

46–47   "I knew perfectly well": *De Stella Nova, TBDOO,* 1:16/Christianson 1964, 122. The full title of the 1573 book, best known as *De Stella Nova,* in which Tycho described the event was (translated into English) *Mathematical Contemplation of Tycho Brahe of Denmark on the New and Never Previously Seen Star First Observed in the Month of November in the Year of Our Lord 1572.*

47   "I doubted no longer": *De Stella Nova, TBDOO* 1:18/Christianson 1964, 123.

47   "Let all philosophers": Ibid.

49   turned the sextant around: *Mechanica,* 84–87.

51   an "oration": The oration is in *TBDOO,* vol. 1.

54   *De Stella Nova:* Ibid.

55   Johannes Kepler wrote: Thoren 1990, 72.

55   Baade concluded: see *Astrophysical Journal* 102 (1945): 309.

56   Radio astronomers: see J. E. Baldwin and D. O. Edge, "Radio Emission from the Remnants of the Supernovae of 1572 and 1604," *The Observatory* 77 (1957): 139ff.; and F. R. Stephenson and D. H. Clark, "The Location of the Supernova of A.D. 1572," *Quarterly Journal of the Royal Astronomical Society* 18 (1977): 340ff.

4. HAVING THE BEST OF SEVERAL UNIVERSES

57   One copy of the invitation still survives: see Thoren 1990, 57, including footnote.

58   "I myself cannot": *TBDOO,* 1:131–32/Thoren 1990, 75.

58   an elegant new quadrant: *Mechanica,* 12–15.

61   "there is something in man," and other quotations from Tycho's lecture: *TBDOO,* 1:163/Thoren 1990, 83.

62   "When I heard": Ibid., 171–72/84.

62   "according to the models," "adapted to the stability": Ibid. 172–73/85–86.

67   "the motions of the planets": Ibid. 172–73/85.

71   Tycho decided he would . . . settle in Basel: *Mechanica,* 108.

74   "I did not want to take": Tycho to Pratensis, mid-February, 1576/ Christianson 1964, 130.

75   "Hear now": Ibid.

5. THE ISLE OF HVEN

77   island of Hven: Tycho Brahe described the island and his palace-observatory in *Mechanica,* 121–40. I am also indebted to John Christianson (2000 and 1961). Christianson 2000 includes biographical information before and after the Uraniborg years and capsule biographies of the assistants who worked for Tycho at Uraniborg and later.

78   "free from the commotion": *TBDOO,* volume 4/Christianson 1961, 120.

79   recounting tales: For the legends of Hven, see Christianson 2000, 112.

80–81   The village's three great fields: Tycho described the topography of Hven and his later improvements to the island in *Mechanica,* 138–39.

80   The map is from a 1588 *Atlas of European Cities.* Tycho later included his own, more accurate map of Hven in *Mechanica,* 138.

82   "to have, enjoy" and "observe the law": *TBDOO,* 14:5/Christianson 1961, 119.

84   Tycho's house plan: The proportions of the structure are described by Thoren 1990, 109.

86   "pipes reaching": *Mechanica,* 129.

87   a ceremony for putting it in place: Tycho described the ceremony in ibid., 130.

87   "consecrated with wines": Ibid.; Tycho reprinted the inscription here.

89   "well-formed" and "over fond of": The horoscope is in *TBDOO,* 1:183–208.

90   he saw an exceptionally bright star: Tycho's description of seeing the comet is in ibid., 4:6.

6. WORLDS APART

Much of the information for this chapter comes from Kepler's "Selbstcharakteristik" of 1597 and other material reprinted as "Heimat" in Schmidt. I am also indebted to Caspar 1948/93.

91 pointed out the bright star with a tail: Kepler told of seeing the comet with his mother in a letter, Kepler to Fabricius, July 4, 1603, *JKGW,* vol. 14, letter 262; and also in his "Selbstcharakteristik," 1597, reprinted as "Heimat" in Schmidt, 220.

92 a noble family: Kepler told about his family's past in a letter, Kepler to Bianchi, Feb. 17, 1619, *JKGW,* vol. 17, letter 827; in Schmidt, 218.

92 candid about his severely dysfunctional family: Kepler's description of his relatives and himself is to be found in Frisch, 8:670–72; for his brothers and sister, 828–29 and 935–36; in Schmidt, 218, 219.

92 "His face" and other quotations: Frisch, 8:670–71; in Schmidt, 218.

96 a careful description: Tycho's study of the comet and his report are in *TBDOO,* vol. 4.

98 "Pseudoprophets" and other quotations from Tycho's report on the comet: *TBDOO,* 4:381–96/Thoren 1990, 130–31. Thoren has used a translation by Christianson.

100–101 He gathered all the observations: Tycho's book about the comet, which he finished in 1588, is in *TBDOO,* vol. 4.

104 "There was nothing I could state": "Selbstcharakteristik"; in Schmidt, 211.

7. A PALACE OBSERVATORY

108 "the winter dining room": The information about the dining room and dining customs, including the menu from another Brahe household, comes from Christianson 2000, 77–78. Christianson cites an early-twentieth-century Scandinavian historian, Troels Frederik Troels-Lund, an expert on daily life in that part of the world in the sixteenth century.

110 Beds were portable: The information about the beds was told me by Henrik Wachtmeister, the present owner of Knutstorps Borg.

110–12 "desks for the collaborators": Tycho's description of the house and garden is in *Mechanica,* 124–32.

112 "one mug after the other": *TBDOO,* 7:327/Christianson 1964, 193.

113 "harmful, uncustomary": Records of Tycho's problems with the islanders, the king's responses to Tycho and the peasants' appeals are in *TBDOO,* vol. 14.

114 "like a mild father": Quoted in Christianson 1964, 33.

115 "workshop for the artisans": *Mechanica,* 139.

121 "people who shun": Ibid., 53.

121 "medium-size azimuth quadrant of brass": Ibid., 16–19.

122 "driven by necessity": Ibid., 144; Tycho described these innovations on 141–144.

Fig. 7.8a "By turning one single screw": Ibid., 143.

8. ADELBERG, HAULBRONN, URANIBORG

126–27 "only those who were hostile" and the other quotations that follow: *JKGW,* 19:328–37; in Schmidt, 221.

129 a giant globe: Information about the globe is in *Mechanica,* 102–5.

129 "by inserting" and other quotations about the globe: Ibid.

131 built the great mural quadrant: Information about the mural quadrant is in ibid., 28–31.

132 "The likeness": Ibid., 30.

134 "with no small difficulty" and "when accidentally": Ibid., 135, 137.

135 Stjerneborg was far from strictly functional: Ibid., 134–37.

138 "great equatorial armillary": Information about the armillary is in ibid., 64–67.

Fig. 8.3 "the two values found": Ibid., 67.

9. CONTRIVING IMMORTALITY

140–42 "Tychonic system of the world": For a detailed description and simple explanation of Tycho's system and its equivalence to Copernicus's, see Kuhn, 202–4. For Tycho's observational campaign to find the parallax of Mars, its connection with the Tychonic system, and Tycho's series of letters about the parallax search, see Gingerich and Voelkel 1998 and Gingerich 1992, 251–56. The latter also tells about the Danish Tycho Brahe scholar John Louis Emil Dreyer.

145 "6:27 P.M.—a meridian altitude": The sequence of observations that took place on March 10 and 11, 1587, is from Gingerich and Voelkel 1998, 17 and 21.

147 A chain of events began: Tycho's description of this incident is translated in Rosen, 39–40. Rosen also reprints excerpts from the eyewitness account of Lange's secretary, Michael Walther (Rosen, 250–53). Christianson's retelling is in Christianson 2000, 89.

148 "four whole handfuls": From Michael Walther's account, reprinted in Rosen, 251, 252.

150 "an evil, scandalous life": Christianson, 126, translating from a book by the nineteenth-century Danish church history scholar Holger Frederik Rørdam.

10. THE UNDERMINING OF THE HUMAN ENDEAVOR

153   "children of poor": See Caspar 1993, 43.

154   "for better and more dignified": *JKGW,* 19:316.

155   "Although [Kölin] once made friends": Biographical material translated in Schmidt, 221.

155–56 "permanent repentance about lost time" and other quotes from Kepler about his work habits: Ibid., 211–13.

158   "I have by degrees": Introduction to *Mysterium; JKGW,* 1:10.

158   the sphere of the stars symbolized Christ: The view of the universe with the Sun, the sphere of stars, and the area between representing Father, Son, and Holy Spirit was a favorite analogy for Kepler all his life. He mentioned it in chapter 2 of *Mysterium,* and much later he used it in book 4 of his *Epitome Astronomiae Copernicanae* (1618–1621); *JKGW,* vol. 7.

158   two formal academic debates: Kepler mentioned his debates in the introduction to *Mysterium; JKGW,* 1:9.

159   "Young Kepler": *JKGW,* 13:4.

160   the third theological year: See Methuen.

161   "tougher than I actually": *Astronomia Nova, JKGW,* 3:108/Caspar 1993, 51.

164–65 Rasmus Pedersen: The Pedersen story is told in detail in Christianson 2000, 332–35.

165–68 As Frobenius described events in his memoirs: Christianson has reprinted Frobenius's account, in English translation, in Christianson 2000, 151–53.

11. YEARS OF DISCONTENT

169   a Rix: Correspondence around the Rix episode appears in *TBDOO,* 6:225–235; Tycho's letter to Wilhelm expressing discontent is on 229.

170   wrote to his friends: For Tycho's letters re the parallax search, see Gingerich and Voelkel 1998, 1, 3–4. Gingerich and Voelkel discuss Tycho's parallax search, its outcome, and his motivation at length.

171   Kepler later examined Tycho's observations of 1582: *JKGW,* 1:439–40; *Astronomia Nova,* Kepler, 1992, Donahue translation, 302.

175–79 arrangement of a marriage: The story of the wedding plans and their failure is told in detail in Christianson 2000, 171–90.

176   courtship and marriage customs: See ibid., 173ff. Christianson cites social historian Troels Frederik Troels-Lund.

180   *Epistolae Astronomicae:* Tycho completed this work in September 1596 and presented copies to the recently crowned King Christian and his chancellor.

180   "This dam and paper-mill": The cornerstone with this inscription
      now resides at Knutstorps Borg.

12. GEOMETRY'S UNIVERSE

183   "foolish little daughter": Kepler repeated the epithet frequently. See
      Gingerich 1973, 290.
183   "nourishing the superstition": *JKGW,* 4:12.
183   "If God gave each animal": Kepler to Mästlin, Dec. 8, 1597.
184   "I inscribed": Introduction to *Mysterium, JKGW,* 1:11.
184   "The delight that I took": Introduction to *Mysterium, JKGW,* 1:13.
Fig. 12.1   Figure 12.1 is a redrawing of Kepler's drawing in *Mysterium.*
186   "I pondered on this subject": Ibid., 1:9.
187   the simple "naturalness" of the cosmos: See Gingerich 1973, 291.
187   "Almost the whole summer": Introduction to *Mysterium, JKGW,*
      1:11.
188   a man created in the image of God could comprehend the logic:
      Kepler voiced this conviction in a letter to Mästlin (Kepler to
      Mästlin, April 19, 1597) and later to von Hohenburg (Kepler to
      von Hohenburg, April 10, 1599). Both are quoted in Holton, 68,
      69.
189   "Finally I came close": Introduction to *Mysterium, JKGW,* 1:11.
190   "And behold, dear reader": Ibid., 13.
191   "Behold, reader, the invention": Ibid.
192   "To see whether this idea": Ibid.
193   "polyhedral theory": The letter was Kepler to Mästlin, Aug. 1595,
      JKGW, vol. 13.
193   "Just as I pledged myself to God": Kepler to Mästlin, Oct. 1595,
      ibid., 40.
195   "lead to the ruin": Mästlin to Kepler, Mar. 9, 1597, ibid., letter 60.
196   "What wonder then": Introduction to *Astronomia Nova, JKGW,*
      vol. 3.
196   "a childish and fateful": Kepler to Duke Friedrich, Feb. 17, 1596,
      JKGW, vol. 8, letter 43.
198   "with very good silk fleece": Papius to Kepler, June 1596, ibid., let-
      ter 45.
198   "set [his] heart on fire": Kepler to Fabricius, Oct. 1, 1602, *JKGW,*
      vol. 14, letter 226.
199   "My assets are such": Kepler to Mästlin, April 9, 1597, *JKGW,* vol.
      13, letter 64.
199   "It is certain": Ibid.

13. DIVINE RIGHT AND EARTHLY MACHINATION

202   "see and learn": Tycho, quoting King Frederick, in Brahe to Pratensis, mid-February 1576/Christianson 1964, 134.

205 (fn)   "most of the larger problems": Stephenson, 1994, 75.

205   "Seldom in history": Gingerich 1973, 292.

205–6   Galileo wrote to Kepler: Galileo to Kepler, Aug. 4, 1597, *JKGW,* vol. 13, letter 73.

206   "would it not be better": Kepler to Galileo, Oct. 13, 1597, ibid., letter 76.

206   "could derive no profit": Praetorius to Herwart von Hohenburg, April 23, 1598, ibid., letter 95.

206   "reviving the Platonic art": Limnäus to Kepler, April 24, 1598, ibid., letter 96.

206   "specialist": Ibid.

207   "The little knowledge . . . I love . . . take care: Kepler to Ursus, Nov. 15, 1595, *JKGW,* vol. 13, letter 26.

207   "most distinguished man": Ursus to Kepler, May 29, 1597, ibid., 124/Rosen, 88.

209   "would have been beheaded": *TBDOO,* 8:7.

212   "I wish he had been there": *Astronomia Nova,* Kepler, Donahue translation, 1992.

14. CONVERGING PATHS

214–15   appeal to King Christian: *TBDOO,* 14:108–11. Dreyer has translated the entire letter, 243–45.

215   Duke Ulrich . . . agreed to intercede: The draft letter to Christian that Duke Ulrich sent Tycho is in *TBDOO,* 14:113, 114.

215   letter went to Lord Chancellor Erik Sparre: Brahe to Sparre, ibid., 119, 120/Thoren 1990, 380.

216   Tycho's coach, drawn by six horses: Christianson 2000, 225. It is not clear precisely when Tycho acquired the horses.

216   "audaciously and not without": Dreyer, 248–52/*TBDOO,* 14: 121–23. Dreyer includes the entire letter.

218   "No doubt the time will come": *TBDOO,* 8:10/Thoren 1990, 381.

218   "Elegy to Denmark": *TBDOO,* 13:101–4. It was copied into the volume containing the observations for 1596 and 1597. Dreyer, 254, gives a description of it but does not print it in its entirety.

220   "The whole German fatherland": Archbishop Elector Ernest of Cologne to King Christian of Denmark, *TBDOO,* 14:140–41/ Thoren 1990, 384.

221–22   "discern double-stars": From Ursus's *De Astronomicis Hypothesibus.*
          Quotations are Jardine's translation; Jardine, 30–36.

222   "The bright glory": Kepler to Ursus, Nov. 15, 1595, *JKGW,* vol. 13,
          letter 26.

223   "the prince of mathematicians": Kepler to Brahe, Dec. 13, 1597,
          ibid., letter 82.

224   "That man [Kepler] has in every way": From "Selbstcharakteristik";
          in Schmidt, 217.

224   "He hurt me with his contempt": *Ibid.*

225   "only better": Kepler to Mästlin, March 15, 1598, *JKGW,* vol. 13,
          letter 89, ll. 180 ff.

225   "Time does not lessen": Kepler to Mästlin, June 1598, ibid., letter
          99.

227   "He who distinguishes himself": Kepler to Johann Georg Brenegger,
          Jan. 17, 1605, ibid., vol. 15, letter 317.

15. CONTACT

231   a letter from . . . Mästlin: Mästlin to Kepler, July 4, 1598, *JKGW,*
          vol. 13, letter 101.

231   reply Tycho had written Kepler: Brahe to Kepler, April 1, 1598, ibid.,
          letter 92.

231   Tycho had sent Mästlin a copy: Tycho's letter to Mästlin, including
          the copy of his letter to Kepler, is Brahe to Mästlin, April 21, 1598,
          JKGW, vol. 13, letter 94.

233   "Why does [Ursus]": Kepler to Brahe, Feb. 19, 1599. This letter was
          not preserved except in a copy that Kepler had made and sent to
          Mästlin.

233   Tycho's response this time: Brahe to Kepler, Dec. 9, 1599, *JKGW,*
          vol. 14, letter 145.

236   He wrote to Edmund Bruce: The letters to Bruce, von Hohenburg,
          and Mästlin were Kepler to Bruce, July 1599, ibid., letter 128; Kepler
          to von Hohenburg, Aug. 6, 1599, letter 130; Kepler to Mästlin, Aug.
          1599, letter 132.

236   Kepler had begun to look to music: A thorough treatment of
          Kepler's first and later efforts to link the planetary orbits with musi-
          cal harmony is Stephenson 1994.

240   "a bird under a bucket": Kepler to Mästlin, Aug. 1599, *JKGW,* vol.
          14, letter 132; and Kepler to von Hohenburg, Aug. 6, 1599, letter
          130.

241   "little paper houses" and "My opinion about Tycho": Kepler to
          Mästlin, Feb. 1599, ibid., vol. 13, letter 113.

16. PRAGUE OPENS HER ARMS

243–44 "perhaps God has acted" and "a splendid": Brahe to Rosenkrantz, Aug. 30, 1599, *TBDOO,* 8:163–66/Thoren 1990, 411.

244 "from what I said": Ibid., 163–66/412.

244 "I saw [the emperor]": Ibid.

245 Tycho's odometer: This was the invention of Peter Jachinow. According to Tycho, it signaled the passing of the miles and portions thereof "by striking distinct sounds with two bells." See Thoren 1990, 205.

245 "the emperor was very favorably": Brahe to Rosenkrantz, Aug. 30, 1599. *TBDOO,* 8:163–66/Thoren 1990, 413.

247 "leave Bohemia": *TBDOO,* 6:273/Thoren 1990, 416.

248 "the women were frightened": *TBDOO,* 8:193, 273/Thoren 1990, 419.

248–49 "No matter what fate": Kepler to Mästlin, Aug. 1599, *JKGW,* vol. 14, letter 132.

249 "I could never torture myself": Ibid.

249 "for in these matters": Mästlin to Kepler, Jan. 15, 1600, *JKGW,* vol. 14, letter 153.

250 "being forced" . . . "desired joint": Brahe to Kepler, Dec. 9, 1599, ibid., letter 145.

251 "as soon as I arrived": Preface to Kepler's "Defense of Tycho against Ursus," Frisch, 1:236–76/Rosen, 330.

251 "You will come": Brahe to Kepler, Jan. 26, 1600, *JKGW,* vol. 14, letter 154.

17. A DYSFUNCTIONAL COLLABORATION

Kepler wrote about his arrival and first weeks at Benatky in letters sent somewhat later to Mästlin and von Hohenburg: Kepler to Mästlin; and Kepler to von Hohenburg, July 12, 1600. *JKGW,* vol. 14, letter 168. See also Gingerich 1973, 294.

253 "saw immediately": Kepler to von Hohenburg, July 12, 1600, *JKGW,* vol. 14, letter 168.

254 "a reigning loneliness": Letter of March 1600; Schmidt, 232.

254 "One day . . . the apogee": Kepler to von Hohenburg, July 12, 1600, *JKGW,* vol. 14, letter 168; in Schmidt, 234.

254 "lofty topics": Brahe to Kepler, April 1, 1598, *JKGW,* vol. 13, letter 92.

254–55 "One of the most important": Kepler to von Hohenburg, July 12, 1600, *JKGW,* vol. 14, letter 168; in Schmidt, 234.

255 "quite a brilliant speculation": Brahe to Mästlin, April 21, 1598, *JKGW,* vol. 13, letter 94.

256   "Tycho has the best observations": Written in March 1600;
        Schmidt, 231.
256   "saw that I possess": Kepler to von Hohenburg, July 12, 1600,
        JKGW, vol. 14, letter 168, in Schmidt, 234.
257   "I thought I would": Ibid.
258   "well-rounded way": Brahe to Kepler, April 1, 1598. *JKGW,* vol. 13,
        letter 92.
258   "Tycho was pleased": Kepler to von Hohenburg, July 12, 1600.
        JKGW, vol. 14, letter 168; in Schmidt, 234.
259   "My greatest worries": From "Selbstcharakteristik"; in Schmidt, 215.
260   "If I don't want": Written in March 1600; Schmidt, 232.
262   "Tycho's house is very cramped": Ibid.
262   "opinion of Brahe's hypotheses": See Rosen, 289, citing *JKGW,*
        14:225.
262   "was eager to know": See Rosen, 289. Rosen cites Frisch, 1:284.
262   "all this is compactly": See Rosen, 290. Rosen cites Frisch, 1:281.
264   a blistering, insulting letter: This was not preserved, only Tycho's
        letter about it, cited below.
264   "find out by a third or fourth": Brahe to Jesensky, Aug. 4, 1600,
        JKGW, vol. 14, letter 161.
264   an apology to Tycho: *JKGW,* vol. 14, letter 162; and *TBDOO,*
        8:305–7.

18. "LET ME NOT SEEM TO HAVE LIVED IN VAIN"

268   "I would not have thought that it is so sweet": Kepler to Mästlin,
        Sept. 9, 1600, *JKGW,* vol. 14, letter 175/Rosen, 281.
269   "with confidence": Brahe to Kepler, Aug. 28, 1600, *JKGW,* vol. 14,
        letter 173.
270   "little professorship": Kepler to Mästlin, Sept. 9, 1600, ibid., letter
        175.
270–71   one last attempt to bluff: Kepler to Brahe, Oct. 17, 1600, ibid., let-
        ter 177.
271   "God let me be bound with Tycho": *JKGW,* 14:203.
272   "Here in Prague": Kepler to Mästlin, Dec. 16, 1600, ibid., letter
        180.
272   "pray for you": Mästlin to Kepler, Oct. 9, 1600, ibid., letter 178.
275   "branded in infamy": Brahe to Rollenhagen, Sept. 26, 1600, *TBDOO*
        8:372/Rosen, 307.
275   "destroying his person": Ibid., 371/307.
278   "still good-natured": Kepler to Mästlin, Feb. 8, 1601, *JKGW,* vol.
        14, letter 183.

278 "choicest" observations: Kepler to Magini, June 1, 1601, ibid., letter 190.

278 "If only I could copy": Kepler to Mästlin, Feb. 8, 1601, ibid., letter 183.

279 "A fever gripped me": *JKGW,* 11:139/Rosen, 322.

279 "Because of this illness": Kepler to Mästlin, Feb. 8, 1601, ibid., letter 183.

279 "rebut even more clearly": Brahe to Kepler, Aug. 28, 1600, *JKGW,* 14:148/Rosen, 299.

279 "Defense of Tycho against Ursus": is in Frisch, 1:236–76, and in English translation in Jardine, 134–207.

280 "If in their geometrical conclusions": From ibid., Frisch, 1:240/ Jardine, 141–42.

280 Barbara wrote: Barbara Kepler to Kepler, May 31, 1601, *JKGW,* vol. 14, letter 188. This is the only one of Barbara's letters that survives. The reason for its survival is that Kepler used the blank parts of the page for astronomical drawings and mathematical calculations. The code Barbara and Johannes Kepler used was deciphered, for Frisch, in the nineteenth century by Otto Struve, director of the Pulkovo Observatory in Russia.

280 "benefactor . . . have more": Eriksen to Kepler, June 13, 1601, ibid., letter 191. Tycho assigned his student Johannes Eriksen to write this letter to Kepler.

282 "so trivial an offense": See Thoren 1990, 469.

283 "Holding his urine" and other quotations about Tycho's death: *TBDOO,* 10:3/Rosen, 312, 313. This account appears at the end of Tycho's collection of observations without indication of who wrote it. The handwriting has been recognized as Kepler's.

284 "although he knew": *Astronomia Nova, JKGW,* 3:89.

285 "The casket": From Kepler's account appended to Tycho's collection of observations; *TBDOO,* 14:233/Thoren 1990, 469, 470.

19. THE BEST OF TIMES

287 "a gathering of nations": Kepler to von Hohenburg, July 12, 1600, *JKGW,* vol. 14, letter 168; in Schmidt, 232.

288 "simply run": Quoted in Caspar 1993, 173.

288 their marriage was not happy: For Johannes and Barbara's married life and the documents that describe it, see ibid., 175–76.

288 "the heart nor the means": See ibid., 176.

289 "weak, annoying": *JKGW,* 19:455.

289 "There was much biting": Ibid., 454.

289 "not much love": Ibid.

290 "unpleasant" and "begrimed" work: *Nova Kepleriana,* 1:18–19.

291 "to think of a lot": From Kepler's "Selbstcharakteristik"; in Schmidt, 212.

293 *Astronomiae Pars Optica:* is in *JKGW,* vol. 2.

295 *A Book Full:* Subtitle of *De Stella Nova,* ibid., vol. 1.

296 "if a pewter dish": Ibid., 1:285.

296 "For what a business": Kepler to von Hohenburg, Dec. 10, 1604.

298 "I consider it a divine decree": *Astronomia Nova, JKGW,* 3:109.

301 "There was nothing I could state": From "Selbstcharakteristik"; in Schmidt, 211.

302 "armed with incredulity": *Astronomia Nova, JKGW,* 3:141.

20. *ASTRONOMIA NOVA*

For a clear, nontechnical discussion of *Astronomia Nova,* see Gingerich 1973, 294–97. For a detailed, chapter-by-chapter discussion, see Stevenson 1987.

306 "If you are wearied": *Astronomia Nova, JKGW,* 3:156.

307 "After divine goodness": Ibid., 178.

309 "The Sun will melt": Ibid., 240.

310 "If one would place a stone": Kepler to Fabricius, Oct. 11, 1605, *JKGW,* 15:358.

310 "see his eyes": *Astronomia Nova,* 243.

317 wrote to his friend David Fabricius: Kepler to Fabricius, July 1603, *JKGW,* 14:410.

317 "Heretofore we have not": *Astronomia Nova,* 310/Stephenson, 103.

318 "I was almost driven to madness": *Astronomia Nova,* 310.

318 "There was nothing": From "Selbstcharakteristik"; in Schmidt, 211.

319 "no longer stand together": Caspar 1993, 135.

21. THE WHEEL OF FORTUNE CREAKS AROUND

321 "not be swayed": Preface to *Astronomia Nova, JKGW,* vol. 3.

323 Galileo's book: *Sidereus Nuncius,* was published in 1610.

324 "I thank you": Galileo to Kepler, Aug. 1610, *JKGW,* 16:327.

325 they viewed Jupiter: Kepler published a report about this study of Jupiter's moons: *Narratio de Jovis Satellitibus* (1611). It was soon reprinted in Florence, where Galileo was living.

325 "I offer you": *Dioptrice, JKGW,* vol. 4.

326 "pulled out Galileo's feathers": Mästlin to Kepler, Sept. 7–17, 1610, *JKGW,* vol. 16, letter 592.

326 *Strena*: The letter is in *JKGW,* vol. 4.

327 "wounded to the depths": Kepler to Scultetus, April 13, 1612. Tobias Scultetus was a friend of Kepler's who was a councillor at the court of Emperor Matthias.

328 "complete the astronomical tables": *JKGW,* 19:123–24.

330 "It makes me heartsick": *Glaubenbekenntnis,* ibid., 12:27.

330 no fewer than eleven candidates: Letter to an anonymous nobleman, dated Oct. 23, 1613. *JKGW,* vol. 17, letter 669.

332 *Nova Stereometria Doliorum Vinariorum: JKGW,* vol. 9.

332 "a little more honorable": Kepler to von Wackhenfels, winter 1618, *JKGW,* vol. 17, letter 783.

333ff accused of witchcraft: Caspar 1993, 241–56, gives a detailed account of the witchcraft trial, based on the acts of the trial and Kepler's letters. Frisch is the source for the trial documents.

22. AN UNLIKELY HARMONY

337 "Since the *Tables* require peace": *JKGW,* 17:254.

339 "sacred frenzy": *Harmonice Mundi,* ibid., 7:290.

340 "O you who by the light": The prayer is at the end of book 5, chapter 9 of *Harmonice Mundi.*

341–43 See note for page 333.

343 "She announced": Frisch, 8:549–50.

344 his library was sealed: Kepler described the sealing of his library in a letter to Paul Gulden, Feb. 7, 1626.

345 filled hundreds of sheets with calculations: It was Owen Gingerich who identified these as preparatory work for the *Tables.*

345–46 "the novelty of my discoveries": Preface to the *Rudolfine Tables, JKGW,* 10:42–43.

346 a book by John Napier: *Mirifici Logarithmorum Canonis Descriptio,* 1614.

347 "on a wagon": Kepler to Matthias Bernegger, Feb. 8, 1627.

350 positions . . . were much more accurate: Accuracy of the *Rudolfine Tables* and 1631 Mercury transit: Gingerich 1973, 305.

23. MEASURING THE SHADOWS

352 to take a presentation copy: Kepler recalled his visit to the court in a letter to Bartsch, Nov. 6, 1629.

353 Kepler was unhappy in Sagan: Kepler to Bernegger, letters dated March 2 and July 22, 1629.

355 "It was meant": Bernegger to Kepler, March 22, 1630.

355   *Somnium:* In *JKGW,* 11:2.

356   lapsed into delirium: The account of Kepler's death comes in part
       from a letter from an unknown scholar named Fischer, in Regens-
       burg, dated January 1631. Several letters about Kepler's death are
       reprinted in Baumgardt, 194–97.

356   "Solely on the merit": Lansius to anonymous, Jan. 24, 1631, *JKGW,*
       vol. 18, letter 1146.

357   "I measured": *JKGW,* 19:393.

BIBLIOGRAPHY

Baumgardt, Carola. *Johannes Kepler: Life and Letters.* New York: Philosophical Library, 1951.

Brahe, Tycho. *Astronomiae Instauratae Mechanica.* Translated into English as *Tycho Brahe's Description of His Instruments and Scientific Work* by Hans Raeder, Elis Strömgren, and Bengt Strömgren. Copenhagen: I Kommission Hos Ejnar Munksgaard, 1946.

Caspar, Max. *Kepler.* Translated and edited by C. Doris Hellman. Original book, in German, was published in Germany in 1948; reissue, with references by Owen Gingerich and bibliographical citations by Gingerich and Alain Segonds, New York: Dover Publications, 1993. Page citations are to the reissued edition.

Caspar, Max, Walther von Dyck, Franz Hammer, and Volker Bialas, eds. *Johannes Kepler Gesammelte Werke.* 22 vols. Munich: Deutsche Forschungsgemeinschaft, and the Bavarian Academy of Sciences, 1937–.

Christianson, John Robert. "The Celestial Palace of Tycho Brahe." *Scientific American* 204 (February 1961): 118–28.

———. *Cloister and Observatory: Herrevad Abbey and Tycho Brahe's Uraniborg.* Ph.D. diss., University of Minnesota, 1964.

———. *On Tycho's Island: Tycho Brahe and His Assistants, 1570–1601.* Cambridge: Cambridge University Press, 2000.

Doebel, Günter. *Johannes Kepler: Er veränderte das Weltbild.* Graz: Verlag Styria, 1983.

Dreyer, John Louis Emil. *Tycho Brahe: A Picture of Scientific Life and Work in the Sixteenth Century.* Edinburgh: Adam and Charles Black, 1890; 2nd edition. New York: Dover, 1963.

————, ed. *Tychonis Brahe Dani Opera Omnia*. 15 vols. Copenhagen: Libraria Gyldendaliana, 1913–29.

Ferguson, Kitty. *Measuring the Universe: Our Historic Quest to Chart the Horizons of Space and Time*. New York: Walker & Company, 1999.

Frisch, Christian, ed. *Joannis Kepleri Astronomi Opera Omnia*. 8 vols. Frankfurt-Erlangen, 1858–1871. This is the source for the witchcraft trial documents, and for many details of Kepler's life.

Gassendi, Pierre. *Tychonis Brahei Vita, Accessit Nicolai Copernici, Georgii Puerbachii et Joannis Regiomontani Vita*. Paris, 1654. Swedish translation, *Tycho Brahe: Mannen och Verket*. Efter Gassendi översatt med kommentar av Wilhelm Norlind. Lund: C. W. K. Gleerup, 1951. The original is extremely rare. Page citations are to the 1951 edition, by way of Thoren's *Lord of Uraniborg*.

Gerlach, Walter, and Martha List. *Johannes Kepler, Dokumente zu Lebenszeit und Lebenswerk*. Munich: Ehrenwirth Verlag, 1971.

Gingerich, Owen. *The Great Copernicus Chase, and Other Adventures in Astronomical History*. Cambridge: Cambridge University Press, 1992.

————. "Johannes Kepler." In *Dictionary of Scientific Biography*, ed. Charles Coulston Gillispie, 7: 289–312. New York: Charles Scribner's Sons, 1973.

Gingerich, Owen, and James R. Voelkel. "Tycho Brahe's Copernican Campaign." *Journal for the History of Astronomy* 29 (February 1998).

Hausenblasová, Jaroslava, and Michal Sronek. *Das Rudolfinische Prag*. Prague: Gallery, 1997.

Holton, Gerald. *Thematic Origins of Scientific Thought: Kepler to Einstein*. Rev. ed. Cambridge, Mass.: Harvard University Press, 1988.

Jardine, Nicholas. *The Birth of History and Philosophy of Science: Kepler's "A Defense of Tycho against Ursus" with Essays on its Provenance and Significance*. Includes a translation of "A Defense of Tycho against Ursus." Cambridge: Cambridge University Press, 1984.

Kepler, Johannes. *Astronomia Nova*. Translated into English as *Johannes Kepler: New Astronomy* by William H. Donahue. Cambridge: Cambridge University Press, 1992.

————. *Harmonice Mundi*. Translated into English as *Five Books of the Harmony of the World* by Eric J. Aiton, A. M. Duncan, and J. V. Field. Philadelphia: 1993.

————. *Johannes Kepler Selbstzeugnisse*. Edited by Franz Hammer; translated into German by Esther Hammer, with commentary by F. Seck. Stuttgart–Bad Cannstatt: TK 1971. Contains Kepler's "Selbstcharakteristik."

————. *Mysterium Cosmographicum*. 1596. *Mysterium* has been translated into English as *Secret of the Universe* by A. M. Duncan. New York: 1981. Page citations are to 1596 edition, via Caspar, et al, 1937–.

————. "Rudolphine Tables: Introduction." Translated by Owen Gingerich and William Walderman. *Quarterly Journal of the Royal Astronomical Society* 13 (1972): 60–73.

————. *Somnium.* Translated into English as *Kepler's Somnium* by Edward Rosen. Madison: University of Wisconsin Press, 1967.

Koestler, Arthur. *The Sleepwalkers: A History of Man's Changing Vision of the Universe.* New York: Penguin/Arkana, 1959. The section about Tycho Brahe and Johannes Kepler, "The Watershed," has been printed separately under that title.

Kuhn, Thomas. *The Copernican Revolution: Planetary Astronomy in the Development of Western Thought.* 1957. Reprint, New York: MJF Books, 1985.

Levenson, Thomas. *Measure for Measure: A Musical History of Science.* New York: Simon & Schuster, 1994.

Methuen, Charlotte. *Kepler's Tübingen,* Brookfield VT: Ashgate Press, 1968.

Morris, Roderick Conway. "Palladio: Reinventing the Classical Past." *International Herald Tribune* (on-line), June 23, 2001.

Nova Kepleriana. This is a series of Kepler documents and research about Kepler printed by the Bavarian Academy of Sciences.

Pippard, Brian. *Science: A Physicist's View.* Unpublished paper.

Porter, Neil A. "Kepler." In *Physicists in Conflict.* Bristol, England and Philadelphia: Institute of Physics Publishing, 1998.

Rosen, Edward. *Three Imperial Mathematicians: Kepler Trapped between Tycho Brahe and Ursus.* New York: Abaris Books, 1986.

Schmidt, Justus. *Johann Kepler, sein Leben in Bildern und eigenen Berichten.* Linz: Rudolf Trauner Verlag, 1970. Contains extensive portions of Kepler's "Selbst-charakteristik" (1597) and other personal writings, under the title "Heimat." Material that I have used from this source has been translated into English with the help of Karoline Krenn of the Universität Salzburg.

Stephenson, Bruce. *Kepler's Physical Astronomy.* New York: Springer-Verlag, 1987.

————. *The Music of the Heavens: Kepler's Harmonic Astronomy.* Princeton, N.J.: Princeton University Press, 1994.

Thoren, Victor. *The Lord of Uraniborg: A Biography of Tycho Brahe.* Cambridge: Cambridge University Press, 1990.

————. "New Light on Tycho's Instruments." *Journal for the History of Astronomy* 4 (1973): 25–45.

Voelkel, James R. *Johannes Kepler and the New Astronomy.* New York/Oxford: Oxford University Press, 1999.

Wilson, Curtis. "How Did Kepler Discover His First Two Laws?" *Scientific American* 226 (March 1972): 92–106.

ART CREDITS

The images on the pages noted have been provided by the following sources.

Archiv der Hauptstadt, Prague: endpapers (View of Prague).
Österreichische Nationalbibliothek: iv and 73.
Rychnov nad Kneznou Castle: iv and 329.
Fredriksborgmuseet, Denmark: 26, 85, color plate (Portrait of Frederik II), and color plate (Portrait of Christian IV).
Mary Lea Shane Archives, Lick Observatory, University of California-Santa Cruz: 63 and 324.
Hessisches Landesmuseum, Kassel: 69.
Landesbildstelle Württemberg, Stuttgart: 94, 127, and 156.
Per Remberg and Johan Runeberg: 109, 135, and color plate (floorplan of Hven).
Schiller-Nationalmuseum, Marbach: 154.
Gavnø Castle, Naestved, Denmark: 173.
Owen Gingerich: 185.
Reprinted from Victor Thoren, *The Lord of Uraniborg,* where it was in turn reprinted from a 1912 Danish book by Vilh Lorenzen: 217.
The Royal Library, Copenhagen: 257.
Yale Ferguson: 212, 267, 276, 284, color plate (Chapel of the Magi), and color plate (Benatky Castle).
Prämonstratenserkloster in Strahov, Prague: 343, color plate (Great Globe), and color plate (Portrait of Tycho Brahe—1598).
New York Public Library (Science, Industry, and Business Library): 354.

Sternwarte Kremsmünster: color plate (Portrait of Johannes Kepler, 1610).

Henrik Wachtmeister: color plate (Knutstorps Borg—modern and sixteenth century).

Astronomiae Instauratae Mechanica, Wandesburgi, 1598: color plate (the great mural quadrant), color plate (elevation drawing for Uraniborg), color plate (Uraniborg garden plan), and color plate (Stjerneborg).

Museum of the Russian Academy of Sciences, St. Petersburg: color plate (wedding medallion portraits of Johannes and Barbara Kepler).

Kunsthistorisches Museum, Wien: color plate (portrait of Rudolf II).

Nationalgalerie, Prague: color plate (Prague riots).

INDEX

In this index, TB is used for Tycho Brahe and JK for Johannes Kepler.